CYBERPHILOSOPHY

METAPHILOSOPHY

METAPHILOSOPHY SERIES IN PHILOSOPHY

Series Editors

Armen T. Marsoobian
Brian J. Huschle

Cyberphilosophy

The Intersection of Philosophy and Computing

EDITED BY

James H. Moor and Terrell Ward Bynum

Blackwell
Publishing

Chapters 1–14 first published in vol. 33 nos. 1/2, and chapters 15–17 first published
in *Metaphilosophy*, vol. 33, no. 3, 2002.

350 Main Street, Malden, MA 02148-5018, USA
108 Cowley Road, Oxford OX4 1 JF, UK
550 Swanston Street, Carlton South, Melbourne, Victoria 3053, Australia
Kurfürstendamm, 57 10707 Berlin, Germany

First published 2002 by Blackwell Publishing Ltd

Library of Congress Cataloguing-in-Publication Data has been applied for

ISBN 1-405-10073-7

A catalogue record for this title is available from the British Library.

For further information on
Blackwell Publishers visit our website:
http://www.blackwellpublishing.com

Contents

Communication and Computers

Computer Ethics

1

INTRODUCTION TO CYBERPHILOSOPHY

JAMES H. MOOR and TERRELL WARD BYNUM

We use the term *cyberphilosophy* broadly to designate the intersection of philosophy and computing. We include not just contemporary interaction of philosophy and computing on the Internet but any interaction between the two disciplines however and whenever it occurs. It may come as a surprise to some that philosophers have been interested in computing for centuries. Thomas Hobbes, for example, suggested that human reasoning is at least in part a kind of reckoning, a manipulation of signs – a view that is still advocated by some researchers in artificial intelligence. René Descartes offered a test to tell the difference between a man and a machine that anticipated the well-known Turing test in the twentieth century. Both Blaise Pascal and Gottfried Leibniz were mathematically inclined philosophers who built early calculating devices.

Cyberphilosophy came into its own during the twentieth century as a result of the formulation of the theory of computing by Alan Turing and others, by analysis of social and ethical implications of computing by Norbert Wiener, and by the development of increasingly sophisticated computers, software, and networks. The possibility of creating intelligent machines, or at least machines that could perform tasks that normally required human intelligence, became a reality that fueled the philosophical debate whether minds should be properly understood as computational devices. The deployment of computing technology in the twentieth century raised conceptual and ethical questions about privacy, property, and power. Information, computationally "greased," flowed far too easily from computer to computer to be contained by established customs and categories. Given these and other developments, philosophers could not ignore the impact of computing even if they wished to do so.

Today, the fact that computing has influenced philosophy of mind and has provided new ethical puzzles and dilemmas is widely acknowledged; but our contention is that computing has had an even greater impact upon philosophy than is often realized. All major areas of philosophy – epistemology, metaphysics, and value theory – have been influenced by computing. Philosophical discussions of issues from aesthetics to science and from political philosophy to the structure of reality have been informed by an understanding and application of computing.

To highlight this trend, we compiled an anthology several years ago,

The Digital Phoenix: How Computers Are Changing Philosophy (Bynum and Moor 1998). We wished to call attention to the many ways in which philosophy was being altered by computing and to underscore the fact that not only philosophical research but also philosophical teaching and professional interaction were being significantly affected by computing. Indeed, philosophical research, teaching, and professional interaction are now harmonized and mutually reinforced in ways that were not possible before computing. Our overall thesis in *The Digital Phoenix* was that computing provides philosophy with new and fertile *subject matter*, *models*, and *methods*. We believe that this trend continues in full stride at the beginning of the present century, and we offer this new anthology of articles on cyberphilosophy to illustrate this fact.

New Subject Matter

New philosophical subject matter that results from computing includes the fundamental notions of computing itself. What is a computation? What is an algorithm? A program? What is information? Is hypercomputation beyond the classical limits of Turing machines possible? Some new subject matter is generated by subfields of computing, such as artificial intelligence, robotics, and artificial life. Is artificial intelligence possible, and by what criteria? What is the relationship between having intelligence and having a body? Can virtual creatures possess understanding? Can computerized creatures that reproduce, evolve, and move about their environment consuming resources be alive? And, of course, further important subject matter concerns the impact of computing on people. What should the proper political, social, and ethical applications of computing technology be?

New Models

Computer models are extremely fruitful for philosophical reflection. Because computational processes are so logically malleable, they provide intellectual clay that can be shaped to formulate ideas, explain events, and test hypotheses. Such models force us to consider intriguing questions: Can the brain be understood as a neural computer? To what extent can evolution be regarded as a natural algorithm? Can the emergence of human communication be explained as a product of recursive computing? Can adjustments to rational belief systems be captured in a computational system? Can creativity be accounted for in terms of partially randomized computational processes? In some cases the models can be instantiated and actually run on computers. Such models provide computer-assisted thought experiments, the results of which might well be unknown and unappreciated without computers. Fecund computer-assisted thought experiments encourage more philosophical inquiry and computer modeling.

New Methods

Computers provide philosophers with new methods that may or may not incorporate modeling. Some of the most widely used computerized tools are teaching applications in logic and in other philosophical subjects. Networked computers also make possible distance learning that combines rapid, continual interaction among teachers and students with sophisticated pedagogical evaluation tools. Computer programs are used for research in philosophy to examine and compare texts in ways that would be prohibitively expensive and time-consuming without computers. Philosophy done with computers has an empirical dimension that distinguishes it from philosophy typically thought of as pure conceptual analysis or synthetic construction. It is not uncommon nowadays to find philosophers and computer scientists forming teams to explore intellectual subjects common to philosophy and computing. And finally, such standard computing tools as e-mail and the Web make possible the rapid, global exchange of professional, philosophical information for purposes like the organization of conferences, the creation of books, the assessment of journal submissions, the establishment of high-quality information banks, and the search for employment opportunities.

Overview of the Contents

Although the primary goal of this collection is to illustrate how computing continues to influence philosophy by providing subject matter, models, and methods, it is important to note that in cyberphilosophy the influence may flow in the other direction as well. Philosophy sometimes has an important catalytic influence on computing. Some of the essays included here illustrate the collaboration of philosophers and computer scientists, and some propose the incorporation of philosophical ideas into computing.

While compiling *The Digital Phonenix*, we had planned a two-hundred-page book, but it mushroomed into a four-hundred-page book, somewhat to the discomfort of our gracious and understanding publisher. Even then we had more material than we could include. This time we have kept closer to our agreed page restrictions. In order to underscore the growth and spread of cyberphilosophy, the articles in this collection are new, unpublished papers, many by established authors but many by younger authors – and all by authors who did not contribute to *The Digital Phoenix*. Our sincere apologies to the many, many authors not included who could easily and rightfully have been included in this cyberphilosophy anthology.

We have organized the articles under a variety of philosophical topics to indicate the wide range of subjects in cyberphilosophy. As the reader might expect, specific papers frequently raise multiple issues that could be classified under different headings. Although, as the articles nicely illustrate, cyberphilosophy often provides novel subject matter, models, and methods for philosophy in general, the essays are organized around five standard philosophical themes – minds, agency, reality, communication, and ethics – to stress as well the relevance of computing to the traditional philosophical enterprise.

Minds and Computers

Computers play a major role in discussions of the nature of minds. Good Old Fashioned Artificial Intelligence (GOFAI), a movement originating in the earliest days of AI, held as one of its major tenets that the key to intelligent behavior is the computerized manipulation of symbols that can stand as representations for external situations. Programmers could organize the symbols so that the representations produced the right results. But if our minds work that way, who or what arranges our symbols? Other computational movements, such as connectionism, eschew the symbolic representational approach to mind. Pete Mandik's "Synthetic Neuroethology" considers a computer simulation and evolution of synthetic animals with neural network controllers that Mandik believes shows that minds are representational in a way that can be explained through evolutionary emergence using artificial-life models. To this end he discusses Framsticks creatures in what he characterizes as "prosthetically enhanced thought experiments concerning the evolvability of mental representations."

John Barker defends folk psychology and argues that his computer model of mind can defend a simulation view of explaining how commonsense interpretation of behavior is accomplished by means of pretense-like operations that deploy the cognitive system's own reasoning capabilities in a disengaged manner. Some critics of this position maintain that our cognitive system would require special mechanisms for such disengaged operation, but Barker argues that his computer model is useful in meeting such philosophical objections.

Marvin Croy sees philosophy of mind becoming increasingly interdisciplinary and argues for the use of computer simulations as a standard part of teaching philosophy of mind courses. Croy claims that when philosophical method becomes more empirical, philosophical inquiry becomes more active and more process oriented. He maintains that computers could strikingly influence how a philosophical field is taught, for they may play an ever more significant role in passing an empirical orientation toward philosophy of mind to the next generation of philosophers.

Phenomenology is often thought of as irrelevant or even hostile to the efforts of artificial intelligence. Anthony Beavers defends the method of phenomenology as setting the right frame of reference for a science of cognition because it makes explicit what belongs to cognition and what belongs in the natural world. His position – contrary to, for example, Hubert Dreyfus's position – gives phenomenology a constructive task in artificial intelligence.

Agency and Computers

"Adaptable Robots," by Gene Korienek and William Uzgalis, is a team effort by a computer scientist and a philosopher to describe how characteristics of adaptive biological systems can be used in designing robots. In particular, they discuss tenets of direct perception, affordances, animacy, design considerations, control principles, and collective computational architecture. The authors claim that the emergent behavior of an experimental, segmented robotic arm that has no internal representations but searches for productive behavioral solutions demonstrates many of these tenets. They argue that it behaves much more like a living thing than conventionally designed robots.

Susan Stuart gives a careful philosophical analysis of agency. She argues that most current situated and embodied systems are too limited to be autonomous agents. A truly autonomous agent has to be active in the world, able to synthesize and order its internal representations from its own point of view. And to do this effectively the agent will have to be embedded.

John Sullins provides advice on how to build a cognitive robotics lab using inexpensive LEGO® MINDSTORMS™ robot kits. This lab provides pedagogical and research opportunities for a number of philosophy courses. The major barriers to participation have been the hardware costs and a lack of technical expertise, but in recent years these barriers have become less of a problem. Sullins presents some ideas about the kinds of experiments that these kits make possible and how to expand beyond the limitations of the basic LEGO® hardware.

Reality and Computers

Philosophers have always aimed to understand the nature of reality and have sought fruitful philosophical approaches that will reveal it. In this spirit Luciano Floridi wishes to define a new field, the philosophy of information. He takes the concept of information as basic and then highlights a broad philosophical domain to be understood in terms of this basic concept. He believes the philosophy of information is a mature discipline with unique topics, original methodologies, and new theories. For him the philosophy of information provides a conceptual transformation of philosophy. Floridi's suggestive proposal if adopted would result in cyberphilosophy becoming a core, coherent, and revolutionary approach to doing philosophy leading to a better understanding of the nature of reality.

In contrast, Randall Dipert critically questions the status of and enthusiasm for cyberphilosophy when he states in his article that "except for logic, computers have not yet noticeably improved the quality of philosophizing, research, or pedagogy." Although he discounts the impact of computing on the philosophy of mind, he does see a glimpse of a new metaphysics involving the concept of information and information trans-formation. In this regard he praises the metaphysical work of philosophers like Eric Steinhart.

Richard Scheines explores the use of computers in representing and calculating causal structures. To do this he discusses directed graphs, a concept well exploited in computer science. Much of this research has enhanced the theory of causation, but work in the application of causal theory to data sets has provided practical results as well.

Communication and Computers

Patrick Grim introduces a series of computer models for a broad range of philosophical issues, including cooperation from a society of egoists, self-reference and paradox in fuzzy logic, a fractal approach to formal systems, and an exploration with models for the emergence of communication. Grim illustrates with many examples how computers enhance philosophi-cal modeling and methods.

Colin Allen, Uri Nodelman, and Edward Zalta discuss a different kind of computer resource in philosophy, the development of the *Stanford Encyclopedia of Philosophy*. Traditional encyclopedias of philosophy have existed for some time, but they drift out of date even when put on CDs. The Web offers a way to provide a continually updated source of informa-tion about philosophy. With an authoritative source of information avail-able, the usual problem of finding too much poor information on the Web is avoided. The members of the Stanford team have developed a dynamic reference work of high quality, and they offer convincing reasons why use of their encyclopedia should be free. They also raise issues of funding and archiving versions of web-based material.

Charles Ess explores the nature of computer-mediated communication. Ess maintains that empirical information can be gathered from different cultures by using computer-mediated communication as a laboratory. He suggests that this information reveals differences in philosophical concepts in such areas as the nature of the self, epistemology, and ethics. For exam-ple, the West may view anonymity on the Net as leading to more open and equal communication, but the East may regard it as undermining face-saving.

Ethics and Computers

Walter Maner considers a variety of possible heuristically guided proce-dures, as opposed to strict algorithms, for use in deciding issues in computer ethics. He proposes the concept of procedural ethics, analo-gous to the concept of procedural epistemology advanced by John Pollock. Maner offers criticisms of such systems, but he argues in the end that this general approach to doing computer ethics is sound and should be explored further.

John Weckert describes some ethical issues of nanotechnology and quantum computing, particularly privacy, and those related to artificial intelligence, implants, and virtual reality. Weckert then examines the controversial claim made by Bill Joy that some research in this field should be halted. Weckert maintains that the case for this claim has not been made. He emphasizes that advantages of research are frequently accompanied by disadvantages and that, in particular, research on the cutting edge of computing raises many difficult ethical questions that need to be pursued.

Jeroen van den Hoven and Gert-Jan Lokhorst enumerate in "Deontic Logic and Computer-Supported Computer Ethics" some of the most basic issues in computer ethics, such as intellectual property, privacy, equal access, and responsibility. They then abstract features of the ethical discourse of these subjects and conclude that deontic, epistemic, action logic (DEAL) might capture this discourse and possibly be embedded in computers. Such logical systems have been studied in detail in philosophical logic, but the authors point out that actually implementing such systems on computers would result in more clarity and understanding. In effect they are proposing a vision that, at least in certain domains and situations, computers can and should be programmed to be ethical.

The Future of Cyberphilosophy
Science and technology have traditionally been sources of inspiration for philosophy. Computer science and computer technology are no exceptions. Philosophy has been taking a computational turn. Cyberphilosophy, the intersection of philosophy and computing, is already a growth area that we believe will become increasingly important in this century. Indeed, in the long run computing may have a greater impact on philosophy than will most other scientific and technological advancements. The rich concepts of computing and the usefulness of computing technology will certainly foster philosophical activity. Computing provides a renewable resource of subject matter, models, and methods. Philosophy in return will play a role in computing, for no scientific or technological enterprise is conducted without a philosophical framework. The intersection of philosophy and computing is not only one to watch but also one in which to participate.

Reference

Bynum, Terrell Ward, and James H. Moor, eds. 2000. *The Digital Phoenix: How Computers Are Changing Philosophy*. Revised edition. Oxford: Blackwell Publishers. First edition published 1998. Italian translation *La fenice digitale: Come i computer stanno cambiando la filosofia*. Milan: Apogeo, 2000.

2

SYNTHETIC NEUROETHOLOGY

PETE MANDIK

Introduction: Diachronic Metaphysics and Synthetic Methodologies

One of the core questions of philosophy, especially philosophy of mind, is the question of representation. What is the relation of the mind to the world such that the mind comes to have representations of the world? Asked in a materialistic vein the question is one of how brains, or physical systems more generally, have representations of the world. The neurophilosophical presumption of this essay is that brains are the relevant physical systems in question.

The question may be asked in both of two versions, one synchronic and the other diachronic. Synchronically speaking, we have reason to believe that brains are sufficiently structured and situated in environments, right now, to traffic in representations of aspects of their environments and beyond. What patterns of structure and activity in the physical universe support, synchronically, the representation of objects, properties, and states of affairs?

Diachronically speaking, we have reason to believe that the universe did not always contain representations. What had to happen, over time, for physical structures to come to bear representational properties? Presumably representations postdate the emergence of organisms but predate the emergence of humans. Thus, the temporal course of the existence of representations in the universe is nested in the temporal course of the existence of biological organisms in the universe. Given this biological contextualization of the problem, a natural supposition is that an adequate account of this temporal course will be heavily imbued with remarks on the contributions of evolution by natural selection – the variable inheritance of fitness – to the emergence of representational structures.

The question of representation is foundational to the cognitive scientific enterprise. Further, certain cognitive scientific methodologies supply tools for answering both the synchronic and diachronic aspects of the question. The methodologies I have in mind are heavily constructivist in the sense of Daniel Dennett's view of cognitive science as reverse engineering (Dennett 1998, 249). The advantage accrued by such an approach helps one cope with what neuroscientist Valentino Braitenberg calls the law of uphill analysis and downhill synthesis: it is more difficult to figure out how Mother Nature contrived to design an organism to accomplish some task than it is to come up with one's own artificial solution (Braitenberg 1984). Once, however, a well-functioning artifact has been created to produce the target phenomena – once the synthesis has been effected – the researcher is often in a better position to tackle the analysis.

The cognitive scientific methodologies of artificial intelligence and computational neuroscience supply the means for a synthetic approach to the synchronic aspects of the problem of representation. The techniques of connectionist modeling and the construction of neural network controllers for autonomous robots allow us to test synthetic hypotheses about the possible neural architectures that will support intelligent behavior. The diachronic questions, however, remain relatively untouched by such techniques. How might the various proposed neural architectures have evolved from other systems? One supposition often made in evolutionary contexts is that each relatively adaptive solution must have as a precursor an incrementally distinct but nonetheless relatively adaptive solution. Thus, for instance, the evolution of eyes with lenses had as incrementally adaptive precursors "pinhole" eyes without lenses, light-sensitive envaginations, and light-sensitive skin patches, respectively (Llinas 2001, 101). For the synthetic approach to the diachronic aspects of cognitive questions we may turn to the techniques of artificial life.

Artificial life involves the application of biological solutions to computational problems and the application of computational solutions to biological problems (Liekens 2001). Typical artificial-life projects involve the computer-simulated evolution of populations of synthetic creatures. Artificial-life approaches to cognitive scientific problems have several features that distinguish them from other synthetic techniques, such as artificial intelligence (including both GOFAI – Good Old Fashioned AI – and connectionist modeling). First, such artificial-life projects involve the modeling of entire organisms. Natural examples of cognitive creatures are extremely complex, and modeling must necessarily simplify. Whereas GOFAI and connectionist approaches typically simplify by focusing on subsystems of agents (by, for instance, creating a program that can convert text to speech), artificial-life approaches focus on relatively simple creatures. Thus, such approaches echo Dennett's "why not the whole iguana?" approach (Dennett 1998, 309).

A second mark of contrast between artificial-intelligence and artificial-life approaches to cognition is the contrast between the relative reliance on designed versus evolved solutions. Artificial life employs the creation of synthetic cognitive systems via evolutionary algorithms. The variable inheritance of fitness may be defined over a finite set of combinatorial elements. Fitness functions may be specified for the evaluation of combinations. Combinations are copied with varying degrees of fidelity, allowing for both the inheritance of fitness and the introduction of mutations into the gene pool. While the specification of the elements and the fitness functions are up to the designer, the evolutionary products are not. In fact, they are often quite surprising to the designer.

Artificial-life approaches to cognition to date have focused on the design and evolution of "minimally cognitive behavior," like obstacle avoidance and food finding (Beer 1990). Such approaches allow for both the evolution of cognitive systems from noncognitive systems and the evolution of complex cognitive systems from comparatively simple cognitive systems. Such projects have involved the evolution of controllers for robots, the evolution of controllers for simulated morphologies, and the co-evolution of both controllers and morphologies.

When Animats Attack: The Revolt against Representation

Animats are synthetic animals, either computer simulated or robotic. Many prominent animat researchers describe their results as posing challenges to the cognitive scientific assumptions that intelligent behavior requires mental representation and computation. Two prominent representatives of this line of attack are Randall Beer and Rodney Brooks. Beer's work concerns computer simulations of insects with neural network controllers capable of guiding them through environments and finding food (Beer 1990). Brooks's early work concerns six-legged mobile robots with control structures that implement a subsumption architecture: a collection of systems each capable of guiding behavior, some of which are able to modulate (subsume) the activity of others (Brooks 1991). Brooks sees the subsumption architecture as avoiding a bottleneck that is introduced by more hierarchical control systems that employ a central control unit that uses representations. According to Brooks, "Representation is the wrong unit of abstraction in building the bulkiest parts of intelligent systems" and "explicit representations and models of the world simply get in the way" (Brooks 1991, 140). Similarly, Beer states of his computer-simulated insect,

> There is no standard sense of the notion of representation by which the artificial insect's nervous system can be said to represent many of the regularities that an external observer's intentional characterization attributes to it. Even the notion of distributed representation which is currently popular in connectionist networks does not really apply here, because it still suggests the existence of an internal representation. . . . The design of the artificial insect's nervous system is simply such that it generally synthesizes behavior that is appropriate to the insect's circumstances. (Beer 1990, 162–63).

Much animat research, thus, construes the animats as *reactive agents*. Reactive agents are able to exhibit a surprising variety of behaviors in spite of their alleged lack of internal representations of their environments. For simple schematic illustrations of the basic principles of reactive agents, consider some of Valentino Braitenberg's thought experiments from his influential book *Vehicles: Experiments in Synthetic Psychology* (Braitenberg 1984). Figure 1 depicts a bird's-eye view of three of Braitenberg's simplest animats (vehicles). A stimulus source in the form of a light is in the upper left-hand corner of the figure. The vehicle on the left has a single sensor with a single excitatory connection to a single motor. Increased sensor activity results in increased motor activity; thus, increased proximity to the stimulus results in higher velocities of the vehicle. The vehicles in the middle and on the right have slightly more complex architectures.

The middle vehicle, with excitatory connections wired in parallel, will turn away from a stimulus because the motor closer to the stimulus will turn faster. If instead this vehicle had inhibitory connections it would move toward the stimulus. The vehicle on the right, with crossed excitatory connections will move toward a stimulus, and with crossed inhibitory connections will move away. Braitenberg's vehicles illustrate how relatively simple architectures can form the basis of coherent, survival-enhancing behavior. By multiplying the number of sensors (for example, light, chemical, temperature, obstacle proximity, and so on) and kinds of connections (excitatory, inhibitory, parallel, and crossed), a single vehicle could be capable of finding nutrients while avoiding toxins and obstacles. These vehicles are reactive agents insofar as their behaviors are driven by reactions to environmental stimuli. In spite of their simplicity, reactive agents are capable of exhibiting minimally cognitive behavior. Creating such agents is the focus of much animat research. This involves situating an animat in an environment in which it must perform tasks that seem (to the researcher) to be conducive to survival. Thus, typical behaviors to model include avoiding obstacles and finding "food."

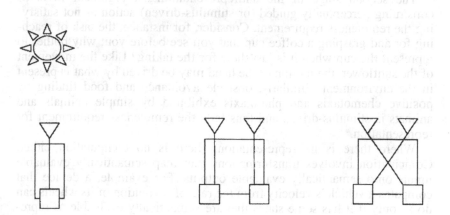

Figure 1. Three Braitenberg vehicles and a stimulus source.

One of the rallying cries of antirepresentational approaches to cognition is that in many, if not all, instances of intelligent behavior, the environment is sufficiently rich that the agent need not represent it, just react to it. Some, such as Brooks, even go so far as to describe the world as its own model (Brooks 1991). Such a view is not limited to those working on animats and includes those working on humans and other natural organisms. Thus, for instance, O'Regan and Noë, in describing the function of human visual perception, describe the environment as being its own representation (O'Regan and Noë 2001). This echoes one of the main themes of Ecological Psychology developed by J. J. Gibson and his followers: the contention that sufficient amounts of information are simply in the environment, just waiting to be picked up by the behaving organism, and thus need not be represented or computed (Gibson 1966).

If there is an argument against positing representations implicit in these sorts of remarks, it seems to be the following two-stage argument. The first stage involves the thesis that representation is required only for the guidance of behaviors concerning things that are somewhat spatially and temporally remote. Remembering what happened last night and a thousand miles away seems to involve some kind of representation or record of these past events. Similarly, plotting a course for the Eiffel tower while one is in Spain cannot be stimulus-driven by the tower itself. Call this the *remoteness requirement* of representation. Haugeland illustrates a similar point by comparing the sun-tracking abilities of a sunflower to an imaginary "supersunflower" (Haugeland 1991). A sunflower is able to track the position of the sun in the sky, but only if the sun is not blocked by a cloud or a building. An imaginary *super* sunflower, in contrast, is able to track the position of the sky by representing its trajectory and can do so even when the sun ducks behind an obstruction.

The second stage of the antirepresentationalist argument involves construing perceptually guided (or stimulus-driven) action as not satisfying the remoteness requirement. Consider, for instance, the task of reaching for and grasping a coffee cup that you see before you: why bother to represent the cup when it is just there for the taking? Like the movement of the sunflower, the motion of the hand may be driven by what is present in the environment. Similarly, obstacle avoidance and food finding by positive chemotaxis and phototaxis exhibited by simple animals and animats is stimulus-driven and thus fails the remoteness requirement for representation.

Where there is no representation, there is no computation either. Computation involves transformations that map semantically evaluable inputs onto semantically evaluable outputs. For example, a device that computes a vehicle's velocity from the rate of revolution of its wheels can do so only if it has some states that are semantically evaluable as representing velocity and other states that are semantically evaluable as representing rate of revolution. Thus, showing that an organism or animat lacks states that satisfy the remoteness requirement suffices to show both that the organism lacks representations and that it lacks computational processes.

The Metaphysics of the Neuron

Views that go hand in hand with the above thinking, therefore, depict representation as an evolutionary Johnny-come-lately. Such views deny that the appearance of organisms with nervous systems that allow them to move toward nutrients and away from toxins coincides with the appearance of representation and computation. I think that this view is mistaken. Even so-called stimulus-driven behavior satisfies the remoteness constraint. Failure to see this is due to a failure to appreciate the relatively enormous distances (both literal and metaphorical) between sensory inputs and motor outputs. It may be appropriate to describe single-celled organisms as simply reacting without representing, but well before the emergence of vertebrates a significant distance opened between stimulus and response. In animals – complex motile organisms – parts on opposite sides of a multicellelar organism need to know what the others are doing. The distances that have to be traversed may be measured in the vast number of cell membranes that must either be crossed or circumnavigated to match appropriate responses to stimuli. Nervous systems are the evolutionary solution to this complex coordination problem. Further, the solution is representational and computational through and through. Transducer neurons pick up environmental information and encode it in representational formats that may be processed by central systems. Among the products of central processing are motor representations that are decoded by motor systems and eventuate in appropriate muscular activity. These representation-manipulating processes are, by definition, computations.

Returning to the questions raised under the heading of diachronic metaphysics, questions arise of how evolution and the incremental emergence of function give rise to the appearance of the first instances of biological representation and computation. Llinas describes the early evolution of nervous systems as follows. In relatively primitive multicellular organisms like sponges, contractile cells respond to direct stimulation. In more evolved organisms like sea anemones, sensory and contractile functions are handled by separate cells, and on some occasions sensory neurons are connected directly to motor cells without any intervening interneurons (Llinas 2001, 11). The evolution of more complex animals is accompanied by the evolution of nervous systems with more interneurons. Thus, the evolution of nervous systems involves an increase in nervous tissue intervening between stimulus and response. This involves an increase in the number of membranes that must be traversed or circumnavigated to relay a signal from one end of the organism to the other, increasing the satisfaction of the remoteness criterion for representation. The evolutionary growth of nervous systems, however, does not involve lengthening straight-line paths from stimulus to response but instead involves complex branching structures. According to Llinas,

> The great advantage provided by such often widely branching interneurons is the ability to "steer with multiple reins." The sensory stimuli activating a few sensory cells may activate a small set of interneurons, which may in turn and, through many spinal segments of connectivity, evoke a complex motor response involving a large number of contractile elements. Through this profusely branching forward connectivity, the animal becomes capable of performing well-defined gross movements that involve many muscles along its body. (Llinas 2001, 81)

I would add that another advantage of the emergence of branching networks of interneurons is that it marks the emergence of processes that not only *relay* information but also *process* information. Thus, the appearance of these networks constitutes the appearance of computation. I explore this theme in further detail in later sections.

The emergence of memory introduces further remoteness (temporal remoteness) and thus clearer satisfaction of the remoteness constraint. Let us define as memory any persistence of encoded information in nervous systems. One plausible neural mechanism for at least short-term memory is recurrence. Figure 2 depicts a simple network with a recurrent connection from the third neuron back to the second neuron. (Longer-term memory may be accomplished by mechanisms that alter the connectivity and/or the connection weights of the network as a function of experience, as in the proposed mechanisms of Hebbian learning or Long Term Potentiation.)

Recurrence, which may serve to implement certain forms of memory, also potentially serves another purpose: the generation of repetitive motion. Most natural forms of locomotion involve the repetitive motion of some limb. Wings are flapped in flying, legs step in walking, and a tail is flagellated in swimming. Further, the muscles driving these repetitive motions need themselves to be driven by repetitive neural signals. One possible source of such signals is a central pattern generator. In the artificial creatures described in this essay, central pattern generators involve recursively connected collections of neurons that sustain oscillatory activ-

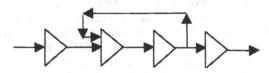

Figure 2. A simple recurrent neural network.

ity. It is possible that the recurrent connections evolved initially to serve as pattern generators for repetitive motions and can be adapted to serve as systems for short-term memory.

The Simulated Evolution of Creatures and Their Neural Networks

The simulations described in the rest of this essay employ the Framsticks artificial-life simulator created by Maciej Komosinski and Szymon Ulatowski (Komosinski 2000; Komosinski 2001). This piece of software allows for the simulation of three-dimensional creatures able to move on land and in water. A sample creature is depicted in figure 3. Creatures are modeled as composed of "sticks": connected finite-length line segments (though typical visualizations depict them as cylinders). Sticks are subjected to simulated physical forces, such as friction, gravity, and buoyancy (in a water environment). Stick creatures may have networks composed of three kinds of neurons: sensors, muscles, and interneurons. Connections between segments may contain either of two kinds of muscles: bending muscles and rotating muscles. The natural pairing of tensors and flexors is simplified into a single muscle that is able to move in either of two opposed directions. Creatures may also be equipped with sensors, of which there are three: smell sensors, touch sensors, and equilibrium sensors. Smell sensors allow the detection of energy sources – food balls – and other creatures, which may be cannibalized under certain conditions. Smell sensors give an increased signal both as a function of proximity of the stimulus and concentration of energy in the stimulus. Touch sensors give an increased signal as a function of proximity to environmental

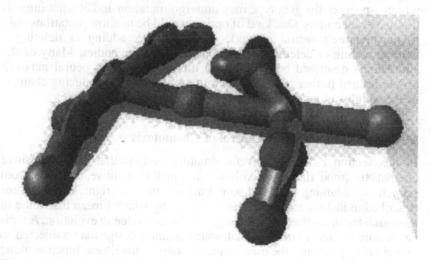

Figure 3. A sample Framsticks creature.

surfaces, such as the ground, obstacles, and barriers. An equilibrium sensor on a stick detects orientation relative to the gravitational field. A horizontal orientation yields a signal of 0, and the signal moves toward -1 or 1 as it is rotated in one of two directions to a vertical position.

Sensors and muscles may be connected by any number of interneurons. Framsticks allows for neural networks of any topology. Neuron states are a sigmoidal function of a weighted sum of inputs. The Framsticks simulator allows for the evolution of creatures via optimization in accordance with weighted combinations of the following fitness criteria: lifespan, horizontal velocity, horizontal distance, vertical position, vertical velocity, body size, and brain size (the number of inputs to neurons in a creature's neural net). Fitness is evaluated by simulating creatures in a virtual world. The structure of a simulated creature is determined by one of the genotypes in the gene pool. When a creature is "born" its genotype is either a clone of a preexisting genotype, a mutation of a preexisting genotype, or a crossover combination of two preexisting genotypes. (There is no sexual differentiation between creatures, but speciation may be introduced by restricting crossovers to relatively similar creatures.) The creature is born into the world with a finite store of consumable and replenishable energy, and the creature dies when either its energy runs out or it suffers a destructive collision with another creature. Fitness of a creature is evaluated by its performance during its lifetime. The calculation of the creature's fitness influences the number of instances of that creature's genotype represented in the genepool. Fitter individuals are more likely to have their genotypes reproduced.

User control of the fitness function is not the only way that evolution may be guided. Framsticks also allows for the user to select separately which aspects of the genotype may undergo mutation and at what intensities and probabilities. One kind of option would be to allow mutations only to the creature's neural networks (for example, adding or deleting a neuron, adding or deleting connection), not to their bodies. Many of the simulations described below allowed mutations only of neural network properties and further restricted the mutations to only introducing changes of the weights of neural inputs and connections.

Modular and Nonmodular Control of Chemotaxis

In this section I describe several creature designs that exhibit positive chemotaxis (food finding). Evolving minimally cognitive creatures from scratch is a daunting task, and so is building them outright. The creatures described in this section were "free formed," by which I mean that some of their architecture is due to user design, and some is due to evolution. A typical feature of free forming is that when creature designs are subjected to evolutionary pressures, the user frequently adjusts the fitness function along the way to "get what he wants." Free forming is a kind of engineering where evolution is used as a forge to temper creature designs. For example, a creature will be designed with a neural architecture such that it seems to the designer that it will map inputs to outputs appropriately to exhibit some target behavior like obstacle avoidance. While specifying the neural connections by hand is relatively trivial, hand coding optimal connection weights is daunting. The evolutionary algorithm, however, can pick up the slack: the designed creatures can be subjected to an evolutionary run whereby only mutations to connection weights are allowed and fitness is defined as amount of horizontal distance achieved in the creatures' lifetime. The creatures are allowed to evolve overnight, and when the experimenter returns to the lab the next day, the connection weights may have been optimized.

One of my goals in this section is to introduce the taxonomies of modular and nonmodular architectures in describing design solutions to the problem of positive chemotaxis. What I mean by modular control of chemotaxis is that the motor system responsible for the maintenance of locomotion is separate from sensory motor systems sensitive to the spatial location of the stimulus.

Creatures that exemplify the modular control of chemotaxic behavior have separate systems for the continuation of forward locomotion and stimulus orientation. A particularly successful food finder is the creature Modular-B depicted in figure 4. Modular-B is a modification of the four-legged food finder created manually and then evolved by Miron Sadziak (see Komsinski 2001).

The neural network controller for Modular-B is composed of two distinct systems, depicted in figure 4. One system is the stimulus-orientation system, which consists of two smell sensors connected to a bending muscle in the creature's torso. When there is a higher degree of activity in the right sensor than in the left, the torso muscle bends the creature toward the right. Likewise, mutatis mutandis, for greater activity in the left sensor. (Crossing the connections from the sensors to the torso muscle would result in an energy avoider. This can be a relatively useful behavior in environments where creatures must avoid each other to avoid destructive collisions.) The other neural system in Modular-B is the locomotion system, which consists of a central pattern generator that drives the leg muscles in a synchronous gait.

Creatures that exhibit nonmodular neural solutions to the problem of chemotaxis do not have distinct circuits for forward locomotion and stimulus detection. For example, consider the water creature Eel2 depicted in figure 5. Eel2's neural network, depicted in figure 6, involves a central pattern generator that drives sinusoidal swimming and is modulated by a single-smell-sensor input. Sufficiently high input signals seize the activity of the central pattern generator. When the input decreases, oscillations in the CPG resume. Eel2 exhibits a pattern of food-finding behavior that is relatively typical of nonmodular food finders. Eel2 swims around in wide curved arcs. Increased activity in the smell sensor results in the arcs

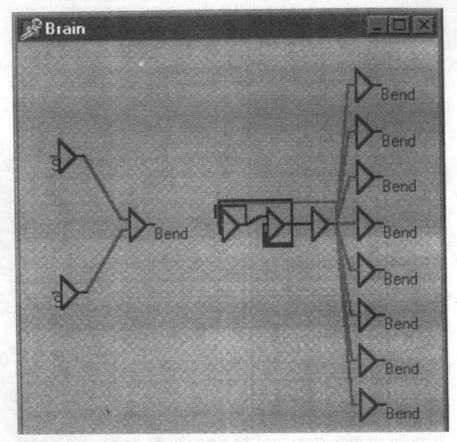

Figure 4. Neural network controller for Modular-B.

becoming tighter, causing the creature to swim in a small circle close
enough to the stimulus source to absorb it. In cases where the stimulus
signal is sufficiently high, Eel2 goes into a "seizure" and stops swimming
altogether. Over evolutionary time Eel2's network became tuned in such a
way that the seizures would occur only if the creature was close enough to
the stimulus source to absorb the food. After a certain amount of food is
absorbed, the seizure ends and Eel2 swims in ever-widening circles,
though still making contact with the food source. Only when the food is
gone will Eel2's body straighten sufficiently to swim away from that location to find the next meal.

The contrast between modular and nonmodular solutions is especially
clear in a comparison between Modular-B and Eel2. Proximity to food
does nothing to modulate Modular-B's central pattern generator and thus

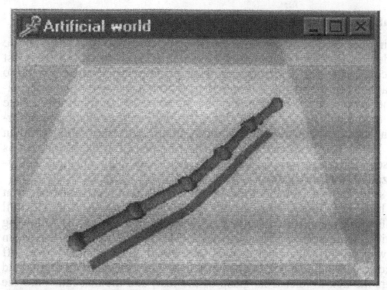

Figure 5. Eel2, a water creature that employs sinusoidal swimming.

does nothing to modulate Modular-B's means of forward locomotion. In contrast, stimulus signals modulate Eel2's locomotion system by changing the waveform of the pattern-generator output. The contrast between food-finding networks of Modular-B and Eel2 also involves important differences in the ways these networks represent and compute information about the creatures' environment. I postpone directly addressing these issues until after laying down further background on the notion of representation.

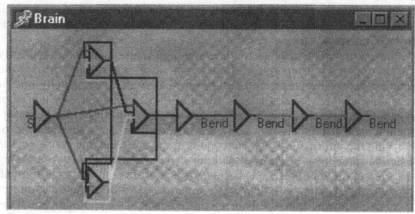

Figure 6. Eel2's neural network.

Mental Representation in a Biological Context

Studying the minimally cognitive behavior of chemotaxis in artificial life provides an excellent opportunity to get at issues concerning the most basic cases of mental representation in an evolutionary context. The basic approach to representation assumed here is both teleological and information theoretic: a representation is a state of an organism (typically, a state of the organism's nervous system) that has the function of carrying information about environmental and bodily states (Dretske, 1988; Milikan 1984; 1993). It is presumed here as a starting point for discussion, but as the discussion progresses, we shall find it necessary to alter the initial characterization of representation.

To see what is useful about this characterization of representation, or at least how it is used, consider the following example. Von Uexkull describes the mental life of a tick as being composed of the following behavioral components (Von Uexkull 1934). A tick clings to leaves on trees waiting for a mammal to pass underneath. The mammal gives off butyric acid, which the tick has detectors for. The detection of butyric acid triggers the tick to release its grasp of the leaf, and the tick falls onto the mammal. Touch sensors detect the appropriate proximity between the tick and the mammal to trigger the tick to run around. The tick continues this running until thermoreceptors detect a high enough temperature to trigger the tick's burrowing response. The tick will then burrow into the skin of the mammal, where it will find the blood it feeds on.

Key adaptive behaviors of the tick are thus driven by mechanisms that function as detectors: they have the functions of carrying information about environmental features. Like ticks, the creatures modeled in Framsticks have chemical receptors and touch sensors, but instead of thermoreceptors, they have equilibrium sensors. Thus, basic abilities to represent environmental features may be built into Framstick creatures at the level of individual transducers. For example, heightened activity in a smell sensor serves to carry information about, and thus represent, concentrations of chemical energy.

Informational approaches to understanding representation are typically cast as opposed to isomorphism-based approaches: approaches that see the relation between representation and represented as fundamentally one of resemblance (see, for example, Cummins 1996). Prototypical instances of nonmental representations can help draw out the opposition. On the information side of the contrast, a ringing doorbell can have the function of carrying the information that someone is at the door and thus represent the fact that someone is at the door without in any significant way resembling the state of affairs of someone being at the door. In contrast, a photograph of someone standing at a door represents that state of affairs in virtue of there being much meaningful resemblance between the photograph and the photographed scene. The geometric arrangement of color and shade in the photograph is isomorphic to the arrangement in the actual scene. For instance, the door is represented as being taller than the person in virtue of the door image being taller than the person image, and so on.

There are, however, other examples that may make the contrast between informational representations and isomorphic representations seem not as sharp. Consider thermometers as devices that carry information about temperature. A mercury thermometer has a column of mercury that changes in height such that the higher the temperature gets, the higher the mercury gets. The column height of, say, three centimeters represents 50° C in the environment in part because that column height is caused by, and thus carries information about, that temperature. But we may also see that isomorphism is at play in the way column heights represent temperatures. The thermometer instantiates an ordered series of physical magnitudes (mercury-column heights) that is isomorphic to an ordered series of physical magnitudes in the environment (temperature). So thermometers constitute a mixed case of informational and isomorphic representations.

Indeed, returning to the case of photographs, they do not represent solely in virtue of isomorphism; causal-cum-informational relations play constitutive roles as well. A photograph may be a picture of Joe and not his identical twin brother Moe in virtue of the causal relations to Joe and in spite of equally resembling both Joe and Moe. The view of information put forward here is thus in keeping with an etymological understanding of information as inFORMation: something carries information about something else in part because of a sharing of form. The boot print carries information about the boot in part because the mud becomes rather literally inFORMed by the boot. The mixture of information and isomorphism illustrated in terms of the thermometer is also exemplified by Framsticks sensor neurons. For example, the range of activity of a sensory neuron is isomorphic to the distance away from the stimulus source. And part of what makes activity in a sensor represent food, and not something else, is that the activity is caused by food, and thus carries information about food, and not something else.

The mere fact that a state of an organism carries information about some environmental feature is insufficient to make that state have the function of carrying that information. Consider people who have fair hair that gets bleached in sunlight. The bleached color of the hair carries information about the light it was exposed to, but the human organism does not use that information, at least not in the significant way in which information about light picked up through the eyes is used. The state of the organism may *carry* the information, but there is no state of the organism that constitutes the *use* of that information. In other words, the information may be *encoded*, but to count as a representation there must be a process by which the organism *decodes* that information. Millikan makes the point in terms of mechanisms of representational production and mechanisms of representational consumption (Millikan 1984; 1993). According to Millikan, in

order for a state of an organism to come to have the function (in the teleological sense of the term) of carrying information, that state has to be naturally selected to carry information. In order to be naturally selected, the
carrying of information must have some effect on the organism's behavior
that contributes to the organism's fitness. And in order for that to happen,
there must be some system of the organism that can channel that information into the modulation of behavior: a system that decodes the encoded
information in a potentially adaptive way.

Representation in Synthetic Neuroethology

With these minimal remarks about key notions of representation in place –
the notions of information, isomorphism, encoding, and decoding – we are
now in a position to see how Framsticks offers a platform for prosthetically
controlled thought experiments concerning mental representation.

I focus here on the way food-finding Framstick creatures utilize
networks with smell-sensor inputs to represent the spatial relations
between the creature and the stimulus source. A return to the discussion of
the contrast between modular and nonmodular solutions to food finding
will be especially useful in this regard. Recall the way that Eel2, a creature
with a single smell sensor, was capable of finding food. Activity in a single
smell sensor represents proximity to the stimulus source: the higher the
activity, the closer the stimulus source. Only one dimension of spatial
information can be encoded in this one-sensor system. Eel2's network is
incapable of representing or computing anything more specific about the
stimulus source, like the direction of the source with respect to the creature. In contrast, the two-sensor system of Modular-B is capable of encoding information about the two-dimensional location with respect to the
stimulus source, information that is decoded by the single turning muscle
in Modular-B's torso. A greater amount of activity in the right sensor than
in the left sensor indicates the stimulus being farther to the right.
Conversely, a greater amount of activity in the left sensor than in the right
sensor indicates the stimulus being farther to the left. The nervous system
of Modular-B is thus capable of representing (in an egocentric reference
frame) the two-dimensional location of a stimulus source. Activity in each
of the individual sensors represents one dimension of spatial information:
near versus far. The network involving two sensors and one turning muscle
is able to compute two dimensions of spatial information: right versus left
as well as near versus far.

While the one-sensor system seems to be at a clear disadvantage to the
two-sensor system, the one-sensor system is not entirely useless to Eel2.
Information about proximity encoded in the sensor is decoded by the
muscular system by making Eel2 swim in circles such that higher amounts
of sensor activity result in smaller-diameter circles. This behavior
increases the likelihood of hitting the food source.

The evolutionary advantages of being able to represent more features of the environment can be demonstrated experimentally with the Framsticks platform. In the experiment described below, creatures with identical morphologies (bodies) but different neural architectures were evolved in conditions in which food was present, fitness was defined only in terms of horizontal distance, and mutations were allowed only in neural weights (defined as input weights to sensors, motors, and interneurons). The morophologies employed were the same as for Modular-B described above. The neural network controllers were significantly more complex than Modular-B's. The creatures below differed from each other in having either two, one, or no smell sensors. Otherwise, the topology of their networks all conformed to the following scheme. There were feed-forward connections from each sensor to each of four interneurons in a hidden layer. An additional four interneurons (a second hidden layer) each had feed-forward and feedback connections to each of the neurons in the first hidden layer. Each of the nine outputs (one torso and eight leg muscles) received feed-forward connections from each of the neurons in the first hidden layer. Creatures began with all neural weights set to zero and were evolved for 200 million steps of the simulation. Each of the three creature architectures (with intitial weights of 0) were subjected to five evolutionary runs. Population statistics were sampled after every 20 million steps of the simulation. The results are graphed in figure 7.

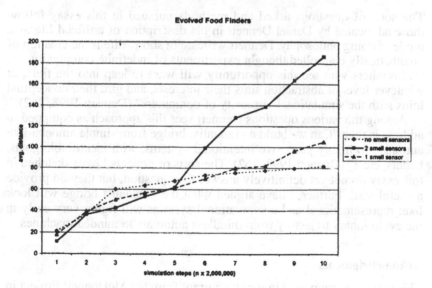

Figure 7. Results of the experiment comparing the evolution of neural weights for land food finders with two, one, and no smell sensors.

Having two smell sensors bestowed a clear advantage over having only one smell sensor, though having one sensor was better than having none at all.

Examining creatures that walk on land to find food allows us to see how simple artificial neural networks are capable of supporting representations of spatial locations of stimuli in one and two dimensions. Switching from land to water renders it is possible to make Framsticks creatures that represent the spatial location of stimuli sources in three dimensions. One way to accomplish this would be with a four-sensor swimming creature. Two of the sensors would be aligned along the creature's horizontal plane and drive a turning muscle that bends the creature along the horizontal plane. The other two sensors would be aligned along the creature's vertical plane and drive a second turning muscle that bends the creature along the vertical plane. Such a system would allow a swimming food finder in relatively deep water to find food sources placed at varying depths. A similar feat could be accomplished with a swimming creature that had only three sensors arranged in a triangle, with, say, one sensor on the top of the creature and the other two on the bottom. The difference between the right and left bottom sensors could be used to drive the horizontal turning muscle. The difference between the top sensor and the sum of the bottom sensors could be used to drive the vertical turning muscle.

Conclusion

The sorts of questions asked and methods pursued in this essay follow those advocated by Daniel Dennett in his description of artificial life as a mode of doing philosophy. Dennett writes: "In short Alife is the creation of prosthetically controlled thought experiments of indefinite complexity. . . . Philosophers who see this opportunity will want to leap into the field, at whatever level of abstraction suits their interests, and gird their conceptual loins with the simulational virtuosity of computers" (Dennett 1998, 262).

Among the various questions Dennett sees this approach as equipped to address is this: "Can we build a gradualist bridge from simple amoeba-like automata to highly purposive intentional systems, with identifiable goals, beliefs, etc.?" (Dennett 1998, 262). The sorts of projects I have sketched in this essay do not yet definitively answer this question, but they do provide a useful start. Further, I have argued what the gradualist bridge will look like: representational and computational systems will figure very early in the evolutionary trajectory from mindless automata to minded machines.

Acknowledgments

This work was supported in part by a grant from the McDonnell Project in Philosophy and the Neurosciences. I thank Eric Steinhart and Shawn Gaston for stimulating discussions of this material.

References

Beer, Randall. 1990. *Intelligence as Adaptive Behavior*. San Diego: Academic Press.

Braitenberg, Valentino. 1984. *Vehicles: Experiments in Synthetic Psychology*. Cambridge, Mass.: MIT Press.

Brooks, Rodney. 1991. "Intelligence Without Representation." *Artificial Intelligence* 47: 139–59.

Cummins, Robert. 1996. *Representations, Targets, and Attitudes*. Cambridge, Mass.: MIT Press.

Dennett, Daniel. 1998. *Brainchildren*. Cambridge, Mass.: MIT Press.

Dretske, Fred. 1988. *Explaining Behavior*. Cambridge, Mass.: MIT Press.

Gibson, J. J. 1966. *The Senses Considered as Perceptual Systems*. Boston: Houghton Mifflin.

Haugeland, John. 1991. "Representational Genera." In *Philosophy and Connectionist Theory*, edited by W. Ramsey and S. Stich. Hillsdale, N.J.: Erlbaum.

Komosinski, Maciej. 2000. "The World of Framsticks: Simulation, Evolution, Interaction." In *Proceedings of 2nd International Conference on Virtual Worlds* (Paris, France). Berlin: Springer-Verlag.

———. 2001. Framsticks website http://www.frams.poznan.pl.

Liekens, Anthony. 2001. Artificial life website http://www.alife.org.

Llinas, Rodolfo. 2001. *I of the Vortex: From Neurons to Self*. Cambridge, Mass.: MIT Press.

Millikan, R. G. 1984. *Language, Thought, and Other Biological Categories*. Cambridge, Mass.: MIT Press.

———. 1993. *White Queen Psychology and Other Essays for Alice*. Cambridge, Mass.: MIT Press.

O'Regan, J. Kevin, and Alva Noë. 2001. "A Sensorimotor Account of Vision and Visual Consciousness." *Behavioral and Brain Sciences* 24, no. 5.

Von Uexkull, Jakob. 1934. *Streifzüge durch die Umwelten von Tieren und Menschen* (a stroll through the worlds of animals and men). Berlin: Springer.

3

COMPUTER MODELING AND THE FATE OF FOLK PSYCHOLOGY

JOHN A. BARKER

Folk psychology, the common-sense view that human behavior is guided by beliefs, desires, and other mental states, is an enigma. Despite its humble origins, this age-old conception of behavior undergirds much of the theorizing of the cognoscenti in such fields as psychology, sociology, anthropology, economics, and philosophy. But what if folk psychology turned out to be a fundamentally flawed empirical theory destined to be displaced by a neurophysiological theory? Mental entities would join phlogiston, caloric, and other assorted debris on the trash heap of discarded posits, taking along myriad theories that presuppose the reality of the mental. The social sciences would undergo drastic revisions, and philosophy as we know it might fade away. Paul Churchland is one philosopher who contemplates such an outcome with enthusiasm. He denigrates folk psychology as an untrustworthy prescientific theory analogous to folk physics and folk biology, and celebrates recent advances in computational neuroscience and connectionist AI as initial steps toward development of a scientifically respectable replacement theory (Churchland 1981, 1988, 1989, 1995, 1998). This theory, he predicts, will eventually give rise to a new common-sense psychology: "The new framework, like any other, will gradually work its way into the general population. In time, it will become the common property of folks generally. It will contribute to, or even constitute, a new folk psychology – one firmly rooted, this time, in an adequate theory of the brain" (Churchland 1995, 323).

In opposition to Churchland's Optimistic Eliminativism, Jerry Fodor advocates Optimistic Preservationism: folk psychology is a fundamentally sound empirical theory that will be vindicated rather than be dislodged by scientific investigations of behavior (Fodor 1983, 1987, 1992a, 1992b, 1993, 1998, 2000). Challenging Churchland's presumption that the common-sense conception is a cultural product, Fodor calls attention to folk psychology's cultural universality and emergence in early childhood, and he hypothesizes that it is an innate intellectual endowment: "Homo sapiens is . . . I suspect, uniquely the species that is born knowing its own mind" (Fodor 1987, 133).

In this essay I discuss recent findings by developmental psychologists that at first blush appear to support Fodor's position. I argue, however, that these appearances are misleading, and that the findings may push both Fodorians and Churchlandians toward Pessimistic Preservationism, the disheartening view that folk psychology is, so to speak, radically wrong but built into our brains. I explore the possibility of rescuing Optimistic Preservationism by dispensing with Fodor's presupposition (shared by Churchland) that folk-psychological interpretation of behavior is based on deployment of a theory. This so-called *theory-theory* has a formidable competitor: the *simulation theory*, which has been advanced by Robert Gordon, Alvin Goldman, Jane Heal, and others (Gordon 1986, 1996; Gordon and Barker 1994; Goldman 1989, 1993; Heal 1986, 1995). According to the simulationist, folk-psychological interpretation of an individual's behavior is accomplished by means of "mental simulations" of the individual's decision-making situation. These pretense-like activities, in which the cognitive system operates in a disengaged, or off-line, manner, allegedly generate reliable predictions and explanations without need of a theory. I discuss the role computational models of rational agents can play in adjudicating the dispute between the simulationist and the theory-theorist, and I describe several models I have constructed for this purpose.

Evidence for Innateness

Simon Baron-Cohen, Uta Frith, Alan Leslie, Josef Perner, Heintz Wimmer, and other developmental psychologists have amassed impressive evidence that children possess innate proclivities toward mentalistic interpretations of behavior (Baron-Cohen 1995; Baron-Cohen, Leslie, and Frith 1985; Wimmer and Perner 1983). By the age of four most developmentally normal children can successfully deploy concepts of belief, desire, and other mental states in explaining and predicting behavior. This "mindreading" capability, as Baron-Cohen calls it, emerges spontaneously, with no need of instruction or superior intellectual endowments. Children with Down syndrome, for example, typically perform in accord with their mental ages. Children afflicted with autism, however, exhibit a strange "mindblindness" that selectively impedes their grasp of the mental. Experimental studies by Baron-Cohen and others reveal that even intelligent autistic adults usually perform poorly on "false belief" tests that are easy for most five-year-olds. Suppose, for example, that Sally puts a marble in a basket and departs, whereupon someone else transfers it to a box. When asked where Sally will look for the marble upon returning, autistic individuals of all ages usually indicate the box, a response characteristic of normal three-year-olds. Numerous other tasks requiring understanding of false belief prove to be beyond the ken of most people with autism, irrespective of intelligence or training.

A Nativistic Theory-Theory

In attempting to explain mindreading capability, Baron-Cohen and Leslie, like Fodor and Churchland, adopt the theory-theory approach (Baron-Cohen 1995; Leslie 1987, 1994). Theory-theorists hypothesize that mindreading abilities are grounded in inferential deployment of theory-like sets of principles. If asked how a child manages to pass the above-mentioned false-belief test, a theory-theorist might reply along the following lines: The child derives the conclusion that Sally will look in the basket by invoking premises about Sally's desires, beliefs, and so on, together with such principles as

(P1) Anyone who wants to find something and believes it to be in a certain place will tend to look for it there.

The premises and principles may be inaccessible to awareness and may be represented in the cognitive system in implicit, nonlinguistic form; they may even be false – as eliminativists are fond of observing, many false theories work fairly well.

Baron-Cohen and Leslie posit the existence of an innate information-processing module, the Theory of Mind Mechanism (ToMM), which interrelates information about volitional and perceptual states and draws conclusions about the full range of mental states: thinking, believing, knowing, deceiving, imagining, pretending, and so forth. On their view, ToMM grounds mindreading ability, and malfunctions in ToMM account for the core impairments associated with autism.[1]

It may appear that the ToMM theory supports Optimistic Preservationism – scientific research already seems to be making strides toward vindicating folk psychology. But we must not take it for granted that evolutionary processes would have tended to provide ToMM with *true* rather than merely *workable* principles. We must ask ourselves a crucial question: How good are Mother Nature's credentials as a theorist? Unfortunately, her theory-construction record is unimpressive. Indeed, as I shall argue, if folk psychology is Mother Nature's own theory of behavior, then there is good reason to fear that it may be radically mistaken.

Empirical theories that match folk psychology with respect to cultural

---

[1] Some developmental psychologists who adopt the theory-theory approach contend that the normal child's innate endowments are insufficiently determinate to account for acquisition of mindreading capability, and that the child learns folk psychology in ways analogous to those in which scientists develop their theories (see, e.g., Gopnik 1996). This view, however, makes it extremely difficult to explain why normal humans in all cultures acquire and retain the same psychology. (To her credit, Gopnik, unlike many proponents of nonnativistic theory-theories, explicitly acknowledges the existence of a culturally universal conception of human behavior and makes a concerted effort to account for its universality, emergence in early childhood, persistence throughout adult life, and continuing influence on the sciences of behavior.)

universality and emergence in early childhood are not easy to find.[2] Perhaps the best candidate is the part of folk physics – call it intuitive physics – that posits the existence of relatively stable objects moving about in various ways under the influence of a variety of forces (pushes, pulls, and so on). Intuitive physics has a conservative, observation-oriented ontology, in that the unobservable objects, forces, trajectories, and so forth that it posits closely resemble observables. Despite this fact, however, key components of the theory are seriously flawed. Consider, for example, the intuitive principle that motion requires force. Notoriously, acceptance of this principle generates numerous difficulties that (thanks to Galileo and Newton) gave rise to a replacement theory containing the unintuitive principle that only change of motion – that is, acceleration or deceleration – requires force. Although this replacement theory has been taught in schools for centuries, it has had little impact on most people's conception of motion. Studies have documented the existence of a deeply ingrained proclivity toward the intuitive principle, a proclivity that not only generates numerous errors in prediction, explanation, and behavior but also significantly impedes the process of learning the replacement theory (McCloskey 1983).

Although intuitive physics may turn out not to be innate, it appears to exemplify what Mother Nature could have been expected to produce were she to have undertaken theory-construction. She would have tended to posit unobservables that resembled observables and would have been apt to make significant errors when venturing much beyond the observable realm. This point can be illustrated by imagining what would have happened had she decided to equip humans with a built-in explanation of the daily movements of celestial bodies. It seems likely that she would have inculcated a theory that these bodies orbit the earth rather than a theory about the earth's rotation. Now it is plausible that she would have found it quite useful for humans to be born knowing their own minds. If folk psychology is indeed her own theory of human behavior, is it apt to be basically true or, like the celestial-orbit theory, fairly workable but fundamentally misguided?

Folk psychology is an intricately structured complex of principles involving a large number of unanalyzable notions that are related to observables in an indirect and collective fashion. The failed projects of the logical behaviorists inadvertently established that these notions cannot be analyzed in terms of relationships to observables. The history of philosophers' unsuccessful attempts to analyze such concepts as knowledge, belief, desire, and purpose, even in terms of other folk-psychological

---

[2] Some Fodorians may be tempted to cite universal grammar as an example of an innate theory that is closely akin to folk psychology. But even if universal grammar can be construed as a theory of some kind, it does not seem to be of the right kind. Fodorians claim that folk psychology is the kind of theory that scientific research could in principle show to be fundamentally false. Such a claim about universal grammar would be highly implausible, to say the least.

notions, provides strong evidence that these concepts acquire their natures at least partly from their roles within the complex of principles. These failures did not stem from vagueness or instability of the notions themselves – both proponents and opponents of proffered analyses could typically reach consensus regarding the effectiveness of numerous counterexamples, something that attests to the existence of a shared fund of intuitively grasped principles.

Clearly, then, folk psychology is a high-level interpretative framework containing a theory-oriented, and hence misconception-prone, ontology. And since many key folk-psychological principles are descriptive only of language-endowed creatures, our distant ancestors would not have constituted fitting subjects for the protracted experimentation that would have been required for discovery and inculcation of principles that correctly explain the behavior of language-using humans. The conclusion seems inescapable that in constructing folk psychology Mother Nature cannot be expected to have hit upon a fundamentally accurate representation of the deeply hidden springs of human behavior. If Fodor is right that folk psychology is an *innate* theory, then there is good reason to fear that Churchland may be right that it is a *false* theory. The bleak prospect of Pessimistic Preservationism looms – folk psychology may turn out to be a misguided theory that will never be displaced, because it is built into our cognitive systems.

Drawing inspiration from Daniel Dennett (1984) and Donald Davidson (1984), we could forestall this disconcerting development by construing folk psychology not as an empirical theory but as a collection of principles that specify what it takes to qualify as a genuine rational agent. We could posit the existence of an innate Rational Agency Mechanism (RAM), which contains representations of the principles and enables the mindreading child to deploy them in a pretense-like fashion. In dealing with the false-belief situation mentioned above, for example, the normal four-year-old could produce the correct prediction by taking "the intentional stance" toward Sally, pretending that she qualifies as a rational agent, as someone who by definition would tend to look for things where they were thought to be. Autistic children (who are known to suffer from pretense-behavior deficits) would fail to produce the correct answer because they could not make proper use of RAM. To help explain how mindless evolutionary processes could have managed to supply RAM with such a remarkably descriptive set of prescriptive principles, we could hypothesize that, via maturation and socialization, the principles themselves tend to shape human behavior, thereby becoming more descriptive of it.[3]

---

[3] This hypothesis might generate legitimate concern about the desirability – and even the rationality – of rationality as defined by the principles in RAM, unless it could be shown that any deleterious or arbitrary principles would tend to be eliminated over time, and that human lives are not now being shaped by a specification of rationality that was appropriate only for our primitive ancestors.

The RAM theory of mindreading capability would have difficulty providing folk psychology with genuine explanatory capability, and it would most likely enjoy no significant advantages over the ToMM theory with respect to simplicity. Furthermore, even though the RAM theory would help dispel Pessimistic Preservationism's air of despair, it would retain so much of its substance that rescuing Optimistic Preservationism would be out of the question. Let us therefore consider a seemingly minor modification in the approach: let us hypothesize that the typical four-year-old is a (reasonably) rational agent who can pretend *to be* Sally and, simply by reasoning as she would, generate the correct prediction. This maneuver, which dispenses with the need to posit representations of folk-psychological principles, exemplifies the guiding strategy of the simulationist.

A Nativistic Simulation Theory

According to the simulation theory, mindreaders do not need a theory about mental states or a conception of rationality to predict and explain behavior. Instead, their cognitive systems employ a pretense-like process to construct "mental simulations" of a target individual's decision-making situation. Their own reasoning capabilities are deployed in a disengaged, or off-line, manner, operating on "pretend" beliefs, desires, feelings, intentions, and so on and yielding generally reliable information about the individual's mental states and associated actions. The culturally universal proclivity toward mentalistic interpretation of behavior is grounded in an innate capacity for mental simulation rather than in an innate theory of mind.[4]

Utilization of the mindreader's own reasoning and decision-making system can generate reliable predictions and explanations of conspecifics because human minds generally function in similar ways. There is no need for explicit or implicit representations of psychological principles – the requisite principles are embodied as operating regularities in the cognitive systems of both the simulating persons and the simulated persons. In the false-belief situation described above, for example, the mindreader's system operates in an off-line manner, forming a desire to find the marble, a belief that it is in the basket, and a rational decision to look there; the simulation culminates in the formation of the on-line belief that Sally will look in the basket. The mindreader need not possess introspective ability or understanding of self; indeed, engaging in primitive forms of simulation grounds acquisition of such ability and understanding. Nor need the

---

[4] The version of simulationism presented here should not be taken to represent all of the many extant versions. For example, this "externalist" simulationism, unlike Heal's "internalist" version, explicates mindreading from a third-person rather than a first-person perspective; unlike Goldman's "introspectivist" version, it entails mindreading not requiring introspective capability; and unlike Gordon's "radical" version, it entails mental state ascriptions not being intrinsically linked to mental simulation, so mindreading could in principle be accomplished without it.

mindreader employ any assumptions about what is transpiring – the entire simulation process can take place automatically, with no conscious awareness on the part of the simulator.[5]

To counter the threat of Pessimistic Preservationism, the simulationist can begin by acknowledging that the mindreader's cognitive system operates *as if* it contained representations of folk-psychological principles, such as

> (P1) Anyone who wants to find something and believes it to be in a certain place will tend to look for it there.

Hence, the system can be described as containing *operative representations* of the principles. The simulationist and the theory-theorist can agree that the mindreader possesses a large fund of such operative representations, a fund constituting what amounts to an *operative theory* of behavior. (To the simulationist, of course, this operative theory is a "virtual" rather than an actual theory, consisting as it does of "virtual" rather than actual representations.) When it comes to dealing with Pessimistic Preservationism, the simulationist enjoys a distinct advantage over the theory-theorist: if simulationism is true, then the mindreader possesses *an operative theory of behavior that cannot be radically misguided*, for its contents reflect the operating regularities of the human cognitive system. There would have been no need for Mother Nature to undertake theory construction. Having produced a basic version of the human cognitive system, she could initiate primitive mindreading activity simply by getting the system to function in a pretense-like manner. And given her penchant for inculcating capacities for mimicry, deception, and pretense in so many of her creatures, she would be apt to adopt this simple strategy to equip humans with an innate device for discovering their minds.

In place of ToMM, the simulationist can posit the existence of what may be called a Disengaged Operation Mechanism (DOM), an innate control mechanism that enables the cognitive system to disengage from normal input-output channels, form and process appropriate mental states, and ascribe them to the relevant individual when engaged operation resumes. Because the system functions in accord with regularities characteristic of mental states, DOM provides the normal child with a "head start" toward

---

[5] While mental simulations can be usefully viewed as inference-like processes that yield "conclusions," the mindreader need not harbor any assumptions to the effect that the two cognitive systems work in similar ways – no analogical inferences are involved. It is the simulation theory itself that contains the assumption that human cognitive systems generally work alike. Hence the theory, if correct, entails and explains simulation-based mindreading as reliably able to generate true beliefs and predictions about behavior, and correct explanations thereof. (Whether or not such mindreading usually provides the mindreader with justified beliefs or with genuine knowledge is a large question that is not being addressed here.)

successful mindreading of conspecifics before any actual representations of principles governing mental goings-on have been acquired. Primitive forms of mindreading lead to development of the standard concepts of belief, desire, and so on – since the concepts of mental states the normal child acquires will be characterized by relationships that mirror the nomic regularities governing the states themselves, there is no need to posit either innate mental-state concepts or concept-acquisition processes that rely on detection of culture-transcending patterns within human behavior. Mindreading practice generates a continuously expanding fund of actual representations of folk-psychological principles, culminating in the accomplished mindreader's consciously accessible conception of human behavior.[6] Concerning autism, abnormalities in DOM's functioning could account for mindreading impairments and the absence of even a basic understanding of human behavior. Furthermore, the fact that DOM operates in a pretense-like fashion could explain the puzzling correlation between such impairments and pretense-behavior deficits linked to autism (Gordon and Barker 1994).

Computational Models of Rational Agents

Stephen Stich and Shawn Nichols (1992) have pointed out that, despite initial appearances, the simulation theory may not enjoy any advantages over the theory-theory with regard to simplicity. According to the simulationist, mindreaders do not need a special database of psychological principles, for off-line reasoning deploys the principles embodied in the cognitive system. According to the theory-theorist, however, mindreaders do not need a special off-line control mechanism, for the principles in the database are deployed via ordinary reasoning processes. Stich and Nichols contend that this dispute cannot be settled until suitable computational models are developed: "While the simulation theorist gets the data base for free, it looks like the theory-theorist gets the 'control mechanism' for free. ... But we don't think either side of this argument can get much more precise until we are presented with up-and-running models to compare. Until then, neither side can gain much advantage by appealing to simplicity" (Stich and Nichols 1992, 53).[7]

---

[6] Although simulational activity (according to this account) initiates the mindreading process in young children, such activity is rapidly and continuously supplemented with formation and use of representations of psychological principles acquired and refined via experience and communication, as well as by additional simulational practice. As many simulationists have emphasized, simulationism need not claim that mental simulation is the only mindreading heuristic, or even the one that most mindreaders predominantly employ.

[7] It may seem that the simulationist can easily win this simplicity competition – even the theory-theorist must acknowledge that normal children already possess a suitable off-line control mechanism, i.e., the mechanism that enables them to engage in ordinary pretense. But some theory-theorists (e.g., Leslie and Baron-Cohen) claim that in order to

In an effort to meet this need, I have developed a software program containing three simple computational models of a rational agent, ProtoThinker (Barker 1998, 1999, 2001). Menu options enable the user to activate and experiment at will with each of the three models. The user interacts with ProtoThinker (PT) through conversation in natural language and can opt to have PT's mental processing displayed in varying degrees of detail to facilitate analysis and evaluation of the modeling strategies.

The three models are the *Core-model*, the *DOM-model*, and the *ToMM-model*, the latter two consisting of the Core-model supplemented with mechanisms that provide PT with mindreading and related capabilities. In all three models PT can be usefully viewed as understanding statements, questions, and requests, forming thoughts and memories, reasoning deductively and inductively, and making rational decisions. In the Core-model, however, PT cannot engage in mindreading, pretending, deceiving, identifying empathetically with others, or understanding metaphorical discourse. As absence of these capabilities in children has been linked to autism, PT can be construed as exhibiting some of the characteristic deficiencies of this affliction. These deficiencies are "remedied," at least to a modest extent, by the mechanisms added to the Core-model to form the DOM-model and the ToMM-model.

The DOM-model, inspired by the simulation theory, contains several disengaged operation mechanisms that give PT limited abilities to mindread, pretend, deceive, empathize, and understand metaphorical statements. They also provide simple forms of additional capabilities that appear to utilize disengaged operations – constructing conditional and indirect proofs, engaging in hypothetico-deductive and analogical reasoning, and predicting grammaticality judgments. The ToMM-model, inspired by the theory-theory, differs from the DOM-model in only one significant respect: PT's mindreading capability is implemented via a separate module called ToMM.[8] As I explain below, the functional distinctness of ToMM from PT's own cognitive operations ensures that the principles employed in mindreading could in principle differ radically from the principles

engage in pretense the child must possess beliefs about pretending, and hence must deploy a theory of mind in the process. Even if this claim could be shown to be predicated on a failure to distinguish carefully between engaging in pretense and understanding pretense, the simulationist would still need to establish that the mechanism that grounds ordinary pretense is also used for mindreading.

[8] The ToMM-model, which was absent from the earliest versions of ProtoThinker, is contained in DOS Version 3.1, Windows Version 4.1, and in all subsequent versions. Windows Version 5.1, which can be downloaded from www.mind.ilstu.edu/research/ai/pt, contains the best implementation of all three models and is the version recommended for experimentation with PT's mindreading capabilities. In the default setting of this version, the DOM-model is active. To activate the ToMM-model, click the "Abilities" button on the main screen and select "Theory-based mindreading." To activate the Core-model, click the "Abilities" button and select "Autism." (For a discussion of computational models of simulational processes from an AI perspective, and arguments for the efficiency of such models as compared with theory-theory models, see Barnden 1995.)

embodied in these operations. In the DOM-model, in contrast, PT's own cognitive operations are themselves deployed in mindreading (albeit in a disengaged manner), and hence the relevant sets of principles are essentially identical.

Currently PT's mindreading abilities are limited to belief prediction, and they do not yet extend to desire prediction, behavior prediction, and so on. To facilitate comparison of the DOM-model and the ToMM-model with respect to simplicity, PT has been given equivalent belief-prediction abilities in each. In both models PT operates as if she possessed actual representations of thousands of folk-psychological principles, such as

(P2) If S believes that X is an A and that every A is a B, then S tends to believe that X is a B.

In other words, PT possesses operative representations of these principles, which taken together constitute the content of her operative theory of belief. By design, PT possesses the same operative theory in both models; but the way in which the theory is implemented differs significantly. In the DOM-model, the content of the theory is directly linked to, and in effect incorporates, numerous principles embodied in PT's own cognitive system. In the ToMM-model, however, the content is wholly contained in the ToMM module, and the principles constituting this content could have differed radically from PT's own principles – ToMM could have been given a theory of belief that was not even roughly descriptive of PT's own cognitive system. Because at present ToMM's theory is highly descriptive of PT's system, the existing ToMM-model accords with Fodorian intuitions. In a suitable Churchlandian model, ToMM's theory would be radically misguided.

Modeling Mindreading Capability

The DOM-model and the ToMM-model utilize the same belief-prediction strategies. It turned out to be necessary to employ a disengaged operation mechanism in the ToMM-model as well as in the DOM-model because reliable belief prediction requires inference replication. To achieve reliability, the mindreader must attend to the propositions constituting the contents of the target individual's known beliefs and utilize these propositions in off-line inferences that roughly accord with the individual's own inferences.[9] To illustrate this, suppose we give a math test to three students,

[9] See Heal 1995 for detailed arguments for the indispensability of inference replication. It may appear that eliminativists could not accept the thesis that reliable belief prediction requires inference replication, as it seems to presuppose realism about beliefs and inferences. But examination of the manner in which inference replication is accomplished in the ToMM-model will indicate that eliminativists need not reject the thesis, even though they would probably choose to word it somewhat differently.

Alice, Betty, and Carol, and try to predict their answers. Alice, let us suppose, is good at math, Betty always divides when she should multiply, and Carol thinks that the word *five* designates the number 9. Those of us who are mathematically challenged will be poor predictors, no matter how expert we are at psychology – we shall have trouble replicating the correct and the incorrect inferences that the students are apt to make.

The following printout of a conversation with PT illustrates the steps in the process by which she attempts to predict beliefs. (For the sake of brevity, only a small portion of PT's mental processing has been reproduced.) Given information about an individual's beliefs, PT simulates possession of the beliefs, tries to replicate the individual's inferences, and, if all goes well, ascribes belief in the conclusions to the individual. To increase the chances that her own off-line inferences really will replicate the individual's inferences, PT always tries to make appropriate adjustments in the level of reasoning competence she uses within the scope of her simulations.

As the conversation begins, PT is told that a certain individual, Brown, believes that Smith is a banker and that every banker is rich. PT is also told that Brown is intelligent. When asked whether Brown believes that Smith is not poor, PT simulates possession of Brown's beliefs and, in view of Brown's high level of reasoning ability, makes a concerted effort to derive the proposition within the context of the simulation.

Conversation 1

Entry: Brown believes that Smith is a banker.
PT's response: I can accept that statement.
Entry: Brown believes that every banker is rich.
PT's response: That's intriguing.
Entry: Brown is intelligent.
PT's response: I did not know that.
Entry: Does Brown believe that Smith is not poor?
(*Simulating subject's beliefs*)
Simulated belief: Smith is a banker.
Simulated belief: Every banker is rich.
(*Assessing subject's reasoning ability; simulating* HIGH *level of reasoning ability*)
Premise: Every banker is rich.
Conclusion: Every banker is wealthy.
Premise: Every banker is rich.
Conclusion: Every banker is not poor.
Premise: Every banker is rich.
Premise: Smith is a banker.
Conclusion: Smith is rich.
Premise: Every banker is wealthy.

Premise: Smith is a banker.
Conclusion: Smith is wealthy.
Premise: Every banker is not poor.
Premise: Smith is a banker.
Conclusion: Smith is not poor.
(*Simulation completed; ascribing relevant belief to subject*)
PT's response: Yes.

By mentally simulating Brown's cognitive situation and attempting to replicate his inferences, PT succeeded in identifying additional beliefs that Brown most likely holds. If we erased PT's thoughts and repeated the conversation without supplying the information that Brown is intelligent, PT would attenuate her reasoning activity within the scope of the simulation; as a result, she would not reach the conclusion that Smith is not poor, and would report that she cannot tell whether Brown believes the proposition. If we asked her whether Brown believes that Smith is wealthy, she would say "Yes," as her off-line reasoning would generate this conclusion with relatively little effort.

Mindreaders must take account of the fact that normally reliable inferences about the beliefs of others, just like normally reliable inferences about world situations, are capable of being defeated by special circumstances. For example, when reasoning leads to tension-generating consequences (for example, conclusions that obviously conflict with existing beliefs), the reasoner usually tries to make adjustments, such as rejecting certain premises or refraining from accepting certain conclusions. PT models this process not only in her own reasoning but also, as the following conversation printout shows, in attempting to replicate the reasoning of others.

As the conversation begins, PT is told that Green believes the following propositions: that Smith is a banker, that every banker is rich, and that Smith is not wealthy. PT is also told that Green is intelligent. When asked whether Green believes that Smith is not poor, PT simulates possession of Green's beliefs, discovers the inconsistency within Green's set of beliefs, and refrains from attributing to Green the belief she would normally ascribe.

Conversation 2

Entry: Green believes that Smith is a banker.
PT's response: I will bear that in mind.
Entry: Green believes that every banker is rich.
PT's response: I can accept that assertion.
Entry: Green believes that Smith is not wealthy.
PT's response: That's nice to know.
Entry: Green is intelligent.

PT's response: I can believe that.
Entry: Does Green believe that Smith is not poor?
(*Simulating subject's beliefs*)
Simulated belief: Smith is a banker.
Simulated belief: Every banker is rich.
Simulated belief: Smith is not wealthy.
(*Assessing subject's reasoning ability; simulating* HIGH *level of reasoning ability*)
Premise: Every banker is rich.
Conclusion: Every banker is wealthy.
Premise: Every banker is rich.
Conclusion: Every banker is not poor.
Premise: Every banker is rich.
Premise: Smith is a banker.
Conclusion: Smith is rich.
Premise: Every banker is wealthy.
Premise: Smith is a banker.
Conclusion: Smith is wealthy.
Premise: Every banker is not poor.
Premise: Smith is a banker.
Conclusion: Smith is not poor.
(*Simulation completed; contradiction detected*)
PT's response: I cannot tell – the individual holds inconsistent beliefs.

Even though PT knows that Smith's not being poor is derivable from the propositions Green believes, she refrains from attributing belief in this conclusion to him because she encounters an inconsistency while attempting to replicate his inferences. In effect, PT assumes that had Green deployed his own reasoning abilities extensively enough to reach this conclusion, he would probably have detected the problem, and his set of beliefs would have undergone some sort of readjustment. Thus, PT models ways in which we try to make allowances not only for variations in people's reasoning abilities but also for their tendencies to try to maintain coherence within their belief systems.

The Simplicity Contest: Which Theory Wins?

Implementation of PT's belief-prediction abilities within the DOM-model required only a small amount of coding. Starting with the Core-model, I added a disengaged operation mechanism, DOM, to enable PT to employ inference replication in such predictions. DOM in effect extracts the propositional contents of metacognitive representations of the form "S believes that P," enables the inference engine to derive conclusions while operating in a disengaged and, if necessary, attenuated manner, searches for inconsistencies in the resultant set of propositions, and incorporates the relevant conclusions into metacognitive representations of the same form if no problems are encountered.

DOM enables PT to predict beliefs without utilizing any actual representations of psychological principles. Because PT's inference engine incorporates hundreds of inference rules that can be combined in numerous ways, DOM automatically endows PT with operative representations of thousands of psychological principles. For example, because PT's inference engine embodies the rule "Given that X is an A and that every A is a B, derive X is a B," DOM provides PT with an operative representation of the psychological principle discussed above:

> (P2) If S believes that X is an A and that every A is a B, then S tends to believe that X is a B.

Indeed, owing to PT's ability to detect conflicts within a set of beliefs, DOM provides her with an operative representation of the following, more sophisticated, version of (P2):

> (P3) If S believes that X is an A and that every A is a B, and S's inferring that X is a B would not tend to generate any conclusions that conflict with other propositions that S believes, then S tends to believe that X is a B.

Moreover, DOM enables PT to take account of variations in reasoning ability and of semantic entailments associated with terms involved in belief statements – for example, the entailment between being rich and not being poor. This has the effect of providing PT with operative representations of numerous predictively powerful principles like the following one, which was implicated in the two sample conversations above:

> (P4) If S believes that X is an A and that every A is a B, and X's being a B semantically entails X's not being a C, and S has a high level of reasoning ability, and S's inferring that X is not a C would not tend to generate any conclusions that conflict with other propositions that S believes, then S tends to believe that X is not a C.

In undertaking the task of constructing a ToMM-model with belief-prediction capabilities equivalent to those in the DOM-model, I encountered a major problem. My initial efforts were guided by the standard theory-theorist view that mindreading is based on deployment of actual representations of psychological principles by means of ordinary reasoning processes. It quickly became apparent, however, that principles like (P3) and (P4) could be deployed only via inference replication, something that would require disengaged use of the model's inference engine.

After numerous unsuccessful attempts to dispense with inference replication, I concluded that ToMM needed its own inference engine, and a disengaged operation mechanism as well. I then realized that providing these would obviate the need for actual representations of psychological principles. A massive fund of operative representations could be supplied in a very compact form – I could provide ToMM with a replica of the DOM-model's inference engine, including its disengaged operation mechanism. This strategy enabled me to construct a ToMM-model that inherited the DOM-model's entire operative theory of belief. When a belief-prediction task is undertaken, the main inference engine sends all information about the target individual to ToMM. ToMM extracts the propositional contents of the individual's beliefs, uses its own inference engine to process them with a sophistication matching that of the DOM-model, and outputs appropriate belief predictions. This design achieves the goal of providing the ToMM-model with belief-prediction capabilities equivalent to those of the DOM-model.

There are several features of the ToMM-model's design that are advantageous from a theory-theory perspective. First, the model accords with the theory-theorist's view that mindreading is grounded in possession of a theory that could, in principle, turn out to be radically misguided. If extensive alterations were made in ToMM without modifying the main inference engine (or vice versa), ToMM would possess an operative theory of belief that was not even roughly descriptive of PT's own cognitive system.

Second, the model accords with Churchland's strategy for dealing with the following vexing problem confronting theory-theorists: "If one's capacity for understanding and predicting the behavior of others derives from one's internal storage of thousands of laws or nomic generalizations, how is it that one is so poor at enunciating the laws on which one's explanatory and predictive prowess depends? It seems to take a trained philosopher to reconstruct them! How is it that children are so skilled at understanding and anticipating the behavior of humans in advance of ever acquiring the complex linguistic skills necessary to express those laws?" (Churchland 1988, 217). To address this problem, Churchland sketches a revised theory-theory that employs the notion of a prototype: "A normal human's understanding of the springs of human action may reside not in a set of stored generalizations about the hidden elements of mind and how they conspire to produce behavior, but rather in one or more prototypes of the deliberative or purposeful process. To understand or explain someone's behavior may be less a matter of deduction from implicit laws, and more a matter of recognitional subsumption of the case at issue under a relevant prototype" (Churchland 1988, 218). Because the ToMM module contains a replica of the core of PT's cognitive system, it functions as a prototype of a rational thinker and dispenses with the need for inferential deployment of representations of folk-psychological principles. The fact that the module contains "virtual" representations of thousands of folk-psychological principles indicates how Churchland's new theory-theory could follow the lead of the simulation theory in construing the sophisticated mindreader's actual representations of such principles and capacities to provide them with verbal garb, as "effects" rather than "causes" of basic mindreading capability.

Finally, the design of the ToMM-model suggests a way in which Churchland could handle a problem confronting his new prototype-based theory-theory. Churchland hypothesizes that the prototypes needed for mindreading can be modeled in "artificial neural networks, networks that mimic some of the more obvious organizational features of the brain" (Churchland 1988, 219). But to model the mindreader's capacity to engage in inference replication, which involves tracking logical relationships among propositions constituting the contents of mental states, such networks would probably have to implement symbol-processing systems, and Churchland generally eschews positing symbol-processing systems to explain human capabilities. By drawing inspiration from the ToMM-model, however, the theory-theorist could envision an artificial neural network that models mindreading via a separate symbol-processing module, a module that is not directly involved in other activities and does not presuppose commitment to realism regarding contentful mental states.[10]

Thus, the ToMM-model not only serves to clarify and test the theory-theory but also suggests promising remedies for some of its problems. The computational resources employed for implementation of belief-prediction capability in this model, however, are hundreds of times greater in volume, and far more complex in structure, than those used for this purpose in the DOM-model. This outcome suggests that the simulation theory will turn out to enjoy a simplicity advantage over the theory-theory. If so, we shall have reason to think that simulationism provides the correct explanation for mindreading capability, and hence that folk psychology is fundamentally sound. Of course, exploration of alternative modeling strategies, utilizing neural-network architecture as well as the classical architecture exemplified by ProtoThinker, must be conducted before a warranted verdict can be reached. Meanwhile, however, the prospects for using simulationism to rescue Optimistic Preservationism appear bright.[11]

References

Astington, J. W., P. L. Harris, and D. R. Olson, eds. 1988. *Developing Theories of Mind*. Cambridge: Cambridge University Press.

[10] See Fodor and Pylyshyn 1988 and Ramsey, Stich, and Garon 1991 for discussion of limitations of artificial neural networks with respect to modeling cognitive-level representations without implementing symbol-processing systems. For a discussion of networks that are capable of implementing symbol-processing systems, see Horgan and Tienson 1996.

[11] I am indebted to Fred Adams, David Anderson, Gary Fuller, Robert Gordon, and James Moor for helpful discussions concerning this essay. An earlier version was presented at the 2000 meeting of the American Philosophical Association, Pacific Division.

Barker, J. A. 1998. *ProtoThinker: A Model of the Mind*. DOS Version 3.1. Belmont, Calif.: Wadsworth Publishing.

———. 1999. *ProtoThinker: A Model of the Mind*. Windows Version 4.1. Belmont, Calif.: Wadsworth Publishing.

———. 2001. *ProtoThinker: A Model of the Mind*. Windows Version 5.1. Available for downloading from www.mind.ilstu.edu/research/ai/pt.

Barnden, J. A. 1995. "Simulative Reasoning, Commonsense Psychology, and Artificial Intelligence." In Davies and Stone 1995b, 247–73.

Baron-Cohen, S. 1995. *Mindblindness: An Essay on Autism and Theory of Mind*. Cambridge, Mass.: MIT Press.

Baron-Cohen, S., A. Leslie, and U. Frith. 1985. "Does the Autistic Child Have a Theory of Mind?" *Cognition* 21: 37–46.

Carruthers, P., and P. K. Smith, eds. 1996. *Theories of Theories of Mind*. Cambridge: Cambridge University Press.

Christensen, S. M., and D. R. Turner, eds. 1993. *Folk Psychology and the Philosophy of Mind*. Hillsdale, N.J.: Lawrence Erlbaum Associates.

Churchland, P. M. 1981. "Eliminative Materialism and the Propositional Attitudes." *Journal of Philosophy* 78: 67–90. Reprinted in Churchland 1989, 1–22.

———. 1988. "Folk Psychology and the Explanation of Human Behavior." *Proceedings of the Aristotelean Society*, suppl. vol. 62: 209–21. Reprinted in Churchland 1989, 111–27.

———. 1989. *A Neurocomputational Perspective: The Nature of Mind and the Structure of Science*. Cambridge, Mass.: MIT Press.

———. 1995. *The Engine of Reason, the Seat of the Soul*. Cambridge, Mass.: MIT Press.

Churchland, P. M., and P. S. Churchland. 1998. *On the Contrary: Critical Essays, 1987–1997*. Cambridge, Mass.: MIT Press.

Clark, A., and P. J. R. Millican. 1996. *Connectionism, Concepts and Folk Psychology: The Legacy of Alan Turing*. Oxford: Oxford University Press.

Davidson, D. 1984. *Inquiries into Truth and Interpretation*. Oxford: Oxford University Press.

Davies, M., and T. Stone, eds. 1995a. *Folk Psychology and the Theory of Mind Debate*. Oxford: Blackwell Publishers.

Davies, M., and T. Stone, eds. 1995b. *Mental Simulation: Evaluations and Applications*. Oxford: Blackwell Publishers.

Dennett, D. 1984. *The Intentional Stance*. Cambridge, Mass.: MIT Press.

Fletcher, G. 1995. *The Scientific Credibility of Folk Psychology*. Mahwah, N.J.: Lawrence Erlbaum Associates.

Fodor, J. 1983. *The Modularity of Mind: An Essay on Faculty Psychology*. Cambridge, Mass.: MIT Press.

———. 1987. *Psychosemantics: The Problem of Meaning in the Philosophy of Mind*. Cambridge, Mass.: MIT Press.

————. 1992a. "A Theory of the Child's Theory of Mind." *Cognition* 44. Reprinted in Davies and Stone 1995b, 109–22.

————. 1992b. *A Theory of Content and Other Essays.* Cambridge, Mass.: MIT Press.

————. 1993. *The Elm and the Expert: Mentalese and Its Semantics.* Cambridge, Mass.: MIT Press.

————. 1998. *In Critical Condition: Polemical Essays on Cognitive Science and the Philosophy of Mind.* Cambridge, Mass.: MIT Press.

————. 2000. *The Mind Doesn't Work That Way: The Scope and Limits of Computational Psychology.* Cambridge, Mass.: MIT Press.

Fodor, J., and Z. Pylyshyn. 1988. "Connectionism and Cognitive Architecture: A Critical Analysis." *Cognition* 28: 3–71.

Goldman, A. 1989. "Interpretation Psychologized." *Mind and Language* 4: 161–85. Reprinted in Davies and Stone 1995a.

————. 1993. "Empathy, Mind, and Morals." *Proceedings and Addresses of the American Philosophical Association* 67. Reprinted in Davies and Stone 1995b, 185–208.

Gopnik, A. 1996. "Theories and Modules: Creation Myths, Developmental Realities, and Neurath's Boat." In Carruthers and Smith 1996, 169–83.

Gopnik, A., and H. M. Wellman. 1995. "Why the Child's Theory of Mind Really Is a Theory." In Davies and Stone 1995a, 232–58.

Gordon, R. M. 1986. "Folk Psychology as Simulation." *Mind and Language* 7: 11–34. Reprinted in Davies and Stone 1995a.

————. 1996. "'Radical' Simulationism." In Carruthers and Smith 1996, 11–21.

Gordon, R. M., and J. A. Barker. 1994. "Autism and the Theory of Mind Debate." In *Philosophical Psychopathology*, edited by G. Graham and G. Stevens, 163–81. Cambridge, Mass.: MIT Press.

Greenwood, J. D., ed. 1991. *The Future of Folk Psychology: Intentionality and Cognitive Science.* Cambridge: Cambridge University Press.

Haselager, W. F. G., ed. 1997. *Cognitive Science and Folk Psychology: The Right Frame of Mind.* London: Sage Publications.

Heal, J. 1986. "Replication and Functionalism." In *Language, Mind, and Logic*, edited by J. Butterfield, 135–50. Cambridge: Cambridge University Press.

————. 1995. "How to Think about Thinking." In Davies and Stone 1995b, 33–52.

Horgan, T., and J. Tienson. 1996. *Connectionism and the Philosophy of Psychology.* Cambridge, Mass.: MIT Press.

Leslie, A. 1987. "Pretense and Representation: The Origins of 'Theory of Mind.'" *Psychological Review* 94: 412–26.

————. 1994. "ToMM, ToBy, and Agency: Core Architecture and Domain Specificity." In *Mapping the Mind: Domain Specificity in Cognition and Culture*, edited by L. Hirschfield and S. Gelman. Cambridge: Cambridge University Press.

McCloskey, M. 1983. "Intuitive Physics." *Scientific American* (Apri):
    122–30.
Ramsey, W., S. Stich, and J. Garon. 1991. "Connectionism, Eliminativism,
    and the Future of Folk Psychology." In Greenwood 1991, 93–119.
Stich, S., and S. Nichols. 1992. "Folk Psychology: Simulation or Tacit
    Theory?" *Mind and Language* 7: 35–71.
———. 1997. "Cognitive Penetrability, Rationality and Restricted
    Simulation." *Mind and Language* 12: 297–326.
Wimmer, H., and J. Perner. 1983. "Beliefs about Beliefs: Representation
    and Constraining Function of Wrong Beliefs in Young Children's
    Understanding of Deception." *Cognition* 13: 103–28.

4

PHILOSOPHY OF MIND, COGNITIVE SCIENCE, AND PEDAGOGICAL TECHNIQUE

MARVIN CROY

That the computational turn provides new methods for research within the philosophy of mind has been well recognized. That these new methods and the knowledge they produce may demand new approaches to teaching the philosophy of mind has been less well recognized. Nevertheless, questions concerning the need for pedagogical change follow innovations in method in any discipline, and the philosophy of mind, given its development in recent decades, is no exception. One clear indication of that development can be found in an announcement of a summer institute funded by the National Endowment for the Humanities in 1981 at the University of Washington.

> Among the most striking changes in recent practice in the Philosophy of Mind is the emergence of a philosophical literature that is sensitive to theoretical and empirical results in Psychology, Artificial Intelligence, Linguistics, and other of the "cognitive sciences." Many philosophers now view their goal less as providing analysis of ordinary language mental expressions than as contributing to the development of empirically and conceptually defensible theories of mental processes.

Changes in the goals of a discipline often require new methods. New aims and methods provide a framework that is fundamental and determines what questions are asked, how questions are framed, and what constitutes plausible answers. Computers have contributed to the quest for empirically and conceptually defensible theories of mind, particularly in the use of simulation as a tool for analysis. Two lines of thought leading to the acceptance of simulation as a method within the philosophy of mind will be outlined here. Particular statements within each of these traditions will be presented, particularly as they emerged during the fruitful decade prior to the 1981 NEH institute. These statements illuminate not only continuing controversies but also certain pedagogical challenges connected to computer simulations and the quest for empirically responsible theories of mind.

From Ordinary Language Analysis to Simulations

Philosophic method has traditionally been associated with and defined in terms of conceptual analysis. Progress in explicating, representing, and evaluating various kinds of logical relations has greatly facilitated conceptual analysis, and its use has been associated with philosophers like Wittgenstein, Wisdom, Austin, and Ryle. Here the emphasis will be on Ryle, as his work founds future methodological developments in the philosophy of mind. Ryle's work is pivotal because it defines conceptual analysis, applies that analysis to mental concepts, and uses the result to critique philosophical theories of mind. His *Concept of Mind* (1949), for example, is rooted in a demolition of dualism. Indeed, Ryle states that the daily use of mental concepts spells doom for Cartesian dualism, for if dualism were true, we could have no knowledge of the mental states of others. Yet our everyday success in employing mental terms constitutes knowledge concerning mental states. Hence, dualism is false. Ryle's position is highly critical of the view that one could know one's private mental life infallibly yet be reduced to "problematic inferences" in respect to the mental life of others. For Ryle, the way out of this muddle is through the explication of the meaning and logical relations of mental concepts, what he termed their "logical geography." In short, philosophers would dissolve such (linguistic) problems by means of conceptual analysis. Ryle sums up his own analyses in precisely these terms.

> I have been examining the logical behaviour of a set of concepts all of which are regularly employed by everyone. The concepts of learning, practice, trying, heeding, pretending, wanting, pondering, arguing, shirking, waiting, seeing and being perturbed are not technical concepts. Everyone has to learn, and does learn, how to use them. Their use by psychologists is not different from their use by novelists, biographers, historians, teachers, magistrates, coastguards, politicians, detectives or men in the street. (Ryle 1949, 319)

Several points should be made explicit about this position. To begin, the logical behavior of concepts depends upon relations of implication, equivalence, consistency and inconsistency, and contradiction. These relationships are explicated by observing the actual use of the concepts. These concepts are not technical terms, nor are they the property of empirical sciences. Finally, there is a presumption concerning the general epistemological adequacy of ordinary mental concepts. This point is crucial. Ryle considers the proper study of psychology and emphatically claims that psychology does not constitute some special realm of science with its own special set of data.

> We find something implausible in the promise of discoveries yet to be made of the hidden causes of our own actions and reactions. We know quite well what caused the farmer to return from the market with his pig unsold. He found that

the prices were lower than he had expected. . . . There are a lot of other sorts of actions, fidgets and utterances, the author of which cannot say what made him produce them. But the actions and reactions which their authors can explain are not in need of an ulterior and disparate kind of explanation. (Ryle 1949, 325)

Ryle holds that this applies even in the fields of perception and memory. It is quite appropriate for psychology to explain perceptual illusions or lapses in memory, but explanations of accurate perceptual judgments are unnecessary. "Let the psychologist tell us why we are deceived; but we can tell ourselves and him why we are not deceived" (Ryle 1949, 326). This view is of much import for questions concerning the relationship between philosophy and science and the significance of the conceptual analysis of ordinary concepts.

A readable elaboration of these views is given by White (1967). To think about a thing is to place it in a certain category, and any categorization commits one to a certain set of inferences. Conceptual analysis explicates this network of inferences, and in doing so "it makes no attempt to dispute, change, or justify the concepts that we employ in thinking about the human mind and human behavior, and which we all use and follow without difficulty in our adult thinking" (White 1967, 15). This reference to our daily and unproblematic use of mental concepts underscores the theme of the epistemological adequacy of such concepts, and it structures the relationship between ordinary mental concepts and those used within psychology. In so doing, it constrains the appropriate aims and methods of psychologists and raises questions about new concepts that emerge in psychological theories. White maintains that even novel and/or technical terms in psychology must be understandable from our ordinary way of thinking, otherwise the relevance of psychology to everyday life would be lost. "Despite the denials of some psychologists, what they label as 'habits,' 'traits,' and 'attitudes' bear very close resemblances to what we ordinarily call by such names. We would not allow them to decide the educational future of our children unless we thought that their 'intelligence tests' measured something very like what we normally think of as intelligence and its manifestations" (White 1967, 12–13).

This position raises questions about the relation between scientific concepts of mind and those employed in everyday life. This issue and that of the relation between science and philosophy will be considered further in a moment. For now, however, let us trace additional links between conceptual analysis and the use of simulations as a method in the philosophy of mind. Todd (1977) supplies one of the most explicit of these links. In a lucid explication of the nature of linguistic analysis, Todd demonstrates the use of flowcharts to improve the technique of conceptual analysis. Since Plato, this analysis has been tied to the search for necessary and jointly sufficient conditions that constitute defining properties of key everyday terms. Todd sees Wittgenstein's notion of family resemblance (that no interesting properties are shared by all instances of a concept) and Quine's thesis of the indeterminacy of translation (which upsets the prin-ciple of synonymy) as deadly for the Platonic enterprise. In consequence, he explores alternative ways of implementing family resemblance within linguistic analysis.

Taking the concept of 'sophist', Todd first proposes the use of varying weights for particular properties of being a sophist. On this view, an individual would be counted as an instance of the concept on the basis of the total score for the sum of its weighted properties. Todd quickly shows, however, that such an approach does not work. The difficulty is that some properties count either more or less in the context of other properties. Todd next offers a "flow chart as an outline for a program whose object would be the simulation of someone's tendency to accept instances of sophistry" (Todd 1977, 275). His flexible and systematic flowchart allows for given properties to be either more important or less, depending upon the presence or absence of other features. Moreover, given the focus on instance acceptance, the distinction between competence models and performance models becomes relevant. A competence model would predict which instances would be accepted, while a performance model would give an account of how people decide whether to accept or reject a potential instance. Todd concludes that philosophers should be concerned with the more demanding task of producing performance models.

> Having gone this far, we cannot deny that philosophy not only rests on empir-ical grounds, but is like the behavioral sciences, at least to the extent that it must ultimately attempt to describe a particular human process, that of recognizing instances under a concept. One might suppose that philosophy merely describes how this is done and leaves it to the behavioral sciences to explain why it is done in the way that it is. However, there is no sharp distinction between the description of such a process and its explanation, and there can be no very extensive description which is not also an explanation. The best way of doing either is to set up a model from which we can derive predictions which can then be compared with observed results. (Todd 1977, 280)

Whatever one thinks of these claims, there is no doubt that Todd contributes to a view in which philosophic inquiry has empirical substance and is intimately connected to the construction of performance models and simulations. In sum, Todd ties together simulations, the empirical dimensions of philosophical analysis, and the way in which logical relationships structure conceptual systems.

Despite his references to simulations and flowcharts, however, Todd never once mentions the role of computing machines. That step is taken, most boldly, by Aaron Sloman (1978). Sloman's *Philosophy and the Computer Revolution* provides a tour de force of the simulation method in AI and its connection to conceptual analysis within philosophy. Sloman is notorious for his claim that within a few years philosophers who could not carry out computer simulations would be considered professionally incompetent. Behind this view stands a portrait of philosophy closely connected to science. This proximity occurs primarily through the interpretative aim of science, an aim concerned not with explanation and prediction but with the discovering of possibilities. Moreover, possibilities are uncovered via the analysis of ordinary concepts. With Ryle, Sloman sees ordinary mental concepts as possessing a privileged epistemological status.

> We have a very rich and subtle collection of concepts for talking about mental states and processes and social interactions. . . . These have evolved over thousands of years, and they are learnt and tested by individuals in the course of putting them to practical use, in interacting with other people, understanding gossip, making sense of behaviour, and even in organizing their own thoughts and actions. All concepts are theory-laden, and the same is true of these concepts. In using them we are unwittingly making use of elaborate theories about language, mind and society. The concepts could not be used so successfully in intricate interpersonal processes if they were not based on substantially true theories. So by analysing these concepts, we may hope to learn a great deal about the human mind and about our own society. (Sloman 1978, 84–85)

Sloman does allow that some sciences devise technical terms that diverge from ordinary meanings, but with respect to mental concepts and the science of psychology he requires that technical terms be formed by uncovering possibilities that lie behind our ordinary mental concepts. Even in physics he sees Einstein's changes in concepts of space and time as the unearthing of flaws in common concepts. Overall, Sloman articulates a view of AI and philosophy that integrates conceptual analysis, concern for ordinary concepts of mind, the explication of possibilities, and simulations. More will be said concerning this integration. For the moment, let us outline a parallel line of thought.

Mental Concepts as Inferred Entities

Another, nearly simultaneous, line of thought also supports the aim of achieving empirically and conceptually defensible theories of mind within philosophy. However, while it emphasizes empirical dimensions of conceptual analysis and acknowledges the potential role that simulations play in that analysis, it repudiates the special status of ordinary concepts of mind and takes more seriously doubts concerning the explanatory role of simulations. Key components of this position are expressed in Fodor's *Psychological Explanation* (1968), which contains two crucial chapters, one on materialism and one on the logic of simulation. Rather than constructing a case for materialism, Fodor argues that materialism may be a logical consequence of one way of viewing mental concepts. This way is quite different from that proffered by Ryle. It recommends that mental concepts be thought of as hypothetical constructs in an empirical theory, much as scientific concepts are. Arguments for this position are developed in Chihara and Fodor (1965), in which the chief target is Wittgenstein's notion of requisite criteria for the application of mental terms. Their criticism runs as follows. Wittgenstein holds that if we can recognize mental state Y on basis of behavior X, then X is either a criterion or a symptom (correlate) of Y. On his view, symptoms necessitate criteria, for one cannot ascertain a symptom of some mental state without being able to recognize the mental state itself. So, having criteria for the application of mental states would be fundamental. Chihara and Fodor resist this conclusion by offering a counter example involving the tracks left by subatomic particles in a cloud chamber. These tracks function neither as criteria nor as symptoms. "Rather, scientists were able to give compelling explanations of the formation of the streaks on the hypothesis that high velocity charged particles were passing through the chamber; on this hypothesis, further predictions were made, tested, and confirmed" (Chihara and Fodor 1965, 291). This view provides an alternative to Wittgenstein's aim of searching for criteria for the application of mental terms and contributes to an alternative view of conceptual analysis.

Once an alternative to the Ryle-Wittgenstein approach is put forward, Fodor (1968) faces questions concerning the methodological standing of simulations. What contribution do simulations make to the attempt to explain human behavior, for example? Is it not possible that a simulation of some cognitive process might produce the same result (behavior) by means entirely different from those that actually occur in human cognition? Fodor's response derives directly from his inferred-entity (hypothetical-construct) view of mental concepts and puts the entire discipline of the philosophy of mind in a new light.

> "When, if ever, would simulating organic behavior by machine count as explaining it?" I have suggested that this question may properly be regarded as a special case of the question: "Under what circumstances does the ability of a theory to account for the relevant observational data make that theory true?" In either case, the answer requires that the counterfactuals predicted by the theory be true and that the theoretical states, entities, and processes posited by the theory exist. On this view, a machine program is simply a way of realizing a psychological theory, and it explains the behavior it simulates only insofar as it satisfies the usual methodological and empirical constraints upon such theories. (Fodor 1968, 145–46)

Two points are worth noting here. First, the constraints referred to will often be onerous. That is, considerable evidence may be demanded in order to support the relevant inferences. A successful simulation of human behavior at best accommodates only a part of that behavior, namely, the part that has been observed thus far. But Fodor insists that "what we are ultimately attempting to simulate when we build a psychological model is not the observed behavior of an organism but rather the behavioral repertoire from which the observed behavior is drawn" (Fodor 1968, 133). Given that we never observe the complete behavioral repertoire, we must make inductive inferences concerning the equivalence of the target and simulation repertoires. Fodor refers to this kind of equivalence as "weak equivalence," and he links it to the truth of relevant counterfactual conditionals that distinguish laws from mere generalizations. Beyond this, there is what Fodor terms "strong equivalence," which requires not only that behavioral repertoires are equivalent but also that these behaviors be produced by the same type of processes. It is strong equivalence that Fodor requires in this context. That is, not only must a simulation produce the correct behavior, it must also produce that behavior in the same way that the simulated organism does. Each of these constitutes a necessary condition for simulations to count as explanations.

Second, Fodor's analysis relies upon success within philosophy of science in explicating the conditions under which theories can be said to be true. This task is yet to be achieved, but that aside, the consequence of Fodor's general perspective is telling. "It seems to me that this is the point at which philosophical questions about psychology merge with more general philosophical questions about scientific theories" (Fodor 1968, 146–47). As Dennett (1978a, 1978b) would later say, the philosophy of mind becomes a branch of the philosophy of science. This perspective provides an avenue for pursuing empirically and conceptually defensible theories of mind outside the Ryle-Wittgenstein framework, and it opens another door to the use of simulations and to artificial intelligence.

Perhaps no one exploits this opportunity at the level of philosophical method more than does Daniel Dennett. While seeing that the aim of psychology is to explain cognition in ways that connect psychological and physiological levels of explanation, Dennett views this task as a kind of engineering AI endeavor. Given the vast difficulties of explaining how the brain actually accomplishes some cognitive function, X, for example, a substitute question is posed: "How could any system (with features A, B, C) possibly accomplish X?" (Dennett 1978a, 111). Being more general and less empirically demanding, this is a more tractable problem, and it is precisely the sort of problem that AI tackles. Nevertheless, the problem can be given more empirical relevance by increasingly instantiating features (such as A, B, C) that function as constraints on any solution. The more that these features represent constraints of actual psychological and physiological processes, the more "psychological reality" one can claim for one's solution. Moreover, Dennett recommends that we think of AI simulations as thought experiments run on computers. Thought experiments provide a kind of conceptual analysis in which the meaning of concepts, their consistency, consequences, and so on, are explored. The same is true within AI. "The questions asked and answered by the thought experiments of AI are about whether or not one can obtain certain sorts of information processing, recognition, inference, control of various sorts, for instance, from certain sorts of designs" (Dennett 1978b, 251). So, Dennett continues a line of thinking that links philosophy of mind with AI and a kind of conceptual analysis unconcerned with ordinary mental concepts and unalarmed at the prospect that these might be improved upon by science.

This view continues the momentum of Kenneth Sayre's pioneering work on recognition. "The value of artificial intelligence to philosophy, rather, is methodological. For artificial intelligence is a form of modelling, and modelling is a technique of conceptual analysis" (Sayre 1979, 140). At a very early stage Sayre saw that an acceptable way, if not the best way, of demonstrating understanding of a mental activity was to produce that activity via a computer program. This perspective is consciously articulated by Zenon Pylyshyn, Roger Shank, and others in Martin Ringle's (1979) *Philosophical Perspectives in Artificial Intelligence*. This work was a major influence on the 1981 NEH institute, which embraced new aims and methods for the philosophy of mind.

Pedagogical Implications

There is one common theme in both lines of thinking described above. This theme is that philosophy, science, and AI are united in the development of empirically and conceptually defensible theories of mind. Both Fodor and Dennett construe the philosophy of mind as a branch of the philosophy of science, but beyond this, their ideas lead in the direction of a view, consistent with that held by Sloman, that the philosophy of mind is rooted in simulation-mediated conceptual analysis and that this analysis forms one end of a continuum comprising empirical inquiry. This conception of a *unified investigation of the mind*, or UIM, has significant consequences not only for research projects but also for pedagogical practice. While opportunities for philosophy to contribute to cognitive-science programs are growing (a point to be addressed later), the main concern here is with courses in the philosophy of mind, traditionally a standard offering within philosophy departments. In what ways should the philosophy of mind be taught to accommodate the principles of UIM?

When the aims and/or methods of a discipline change, change in pedagogy usually follows. This can occur with respect both to content (what is taught) and to pedagogical technique (how it is taught). Within the subject matter's content, a distinction can be made between substantive issues and methodological issues. There is little doubt that content within philosophy of mind courses has changed in line with emerging substantive and methodological issues. Central to these courses are the writings of Turing, Fodor, Dennett, Searle, the Churchlands, Haugeland, Boden, Sloman, and others who assess computational concepts and approaches. There has been much less change, however, in respect to pedagogical technique. There appears to be little recognized need to move beyond a teaching style rooted mostly in lectures and discussion.[1] Few textbooks provide exercises in carrying out simulations or designing cognitive-system components. Nevertheless, if the methods of this field have indeed changed, instructional materials should do more than just describe those changes. They should provide training in using the methods to achieve stated aims. Traditionally, this has occurred by modeling logical and conceptual analysis for students, providing practice that develops such skills, and expecting students to perform analyses on their own. Similar activities may now be relevant for teaching contemporary philosophy of mind, but if so, what form should they take?

It should not be expected that this question can be answered in the abstract. Differences in departmental orientation, student level, curricular context, and so on justify differences in approach. The question will be addressed here in the context of conditions that apply within my own department of philosophy, housed (as is typical) within a college of arts and sciences. After some description of our philosophy of mind course and its history, I shall describe some plans for its continued evolution with respect to emerging instructional activities.

Contemporary philosophy of mind has attracted the attention of other disciplines not only with respect to research but also with respect to the curriculum. In my university's case, this has occurred primarily in the departments of psychology and biology, and to a lesser extent with programs in computer science. Philosophy of mind is now a popular elective within two interdisciplinary undergraduate programs, cognitive science and neuroscience. Our philosophy of mind course attracts as many psychology and as many biology majors as it does philosophy students. Philosophy majors, who once outnumbered the other students (most of them in the humanities), now constitute approximately one-third of the class. This composition shapes both student responses to the subject matter and the dynamics of student interaction. Students are initially intrigued by the possibility that the study of the mind is a unified endeavor, integrating philosophy, science, and AI. However, substantive issues (such as the nature of consciousness) soon become contentions. Students then retreat to the frameworks and methods of their own discipline, which is seen as exclusively addressing the genuinely important questions concerning the mind. The effect on class discussions is predictable. When assessing an issue (such as the prospects for machine intelligence) philosophy students often exhibit near surgical precision in linguistic analysis, while science and engineering students parade the most recent successes in animal or machine performance. A battle of "undercut by analysis of meaning" versus "trump by latest discovery" ensues. This tone becomes less

---

[1] Surprisingly, there is not one article published in *Teaching Philosophy* that explores innovations in teaching the philosophy of mind. Some information of general relevance can be found in the Newsletters published by the American Philosophical Association.

contentious once the limitations of any one perspective emerge and students seriously take up the task of integrating empirical and analytical activities. More challenging from a pedagogical point of view is the fact that students from different disciplines find particular aspects of UIM troublesome. Below, a sketch of these issues and student reservations is followed by a description of two pedagogical responses, one that examines the coherence of UIM and one that elaborates relevant activities that move beyond classroom lectures and discussion.

Student Characteristics

Students from different disciplines bring perspectives that structure how issues in the philosophy of mind are perceived. Philosophy students come to this course with (sometimes considerable) training in logical and conceptual analysis of the kind rooted in Platonic dialogues. They generally approve of the systematic approach exemplified by Todd but are unclear about its relation to empirical inquiry. Moreover, these students take the lead in articulating the possibility that UIM might be too narrow and might ignore important philosophical issues. One example involves the obvious issue of qualia and whether functionalism misses an important feature of the mind, but others involve metaphysical issues, such as the interaction of mind and body.

Science students, on the other hand, are often bothered by the potential discrepancy between simulations and explanations. While recognizing the plausibility of arguments designed to alleviate these concerns, they are for the most part not comforted by them. In fact, science students offer the best candidates for simulations that have little explanatory value. These students arrive with beliefs in the value of explanation and prediction within science, and although they understand the role of thought experiments, they see the explication of new possibilities in a subservient role. Finally, they are troubled by any de-emphasis of the role of testability.

As for computer-science students, they come to the course with training in top-down design and iterative procedures that use feedback to improve performance steadily and control the behavior of complex systems. They immediately grasp the challenge of the homunculus problem. In fact, they have first-hand experience with a practical version (in which one mistakenly assumes that a component's function can be implemented with existing, well-understood resources). They have doubts about how we can know when to stop the top-down decomposition process and declare that some function can be carried out by the "wetware" (which is also to say that they have little confidence in bottom-up development). They are sometimes more inclined toward a neural or artificial network approach to cognition and question the usefulness of propositional attitudes. Also of interest is the fact that the predominant mode of evaluation differs for these student groups. Philosophy students are normally expected to produce written critiques and arguments, while science students face objective tests and write succinct research reports, and computer-science students create, implement, and document algorithmic designs. Hence, the skills of these students often differ.

These descriptions constitute something of a caricature, but they reveal important aspects of UIM that students in the philosophy of mind course find questionable. Let us consider two pedagogical strategies for coping with these concerns. One approach is to meet student questions head-on and, while exposing inconsistencies and potential weaknesses, to provide as clear an argument for UIM as is possible. The aim is not to coerce acceptance but to direct criticisms toward a strong, undiluted case. A second approach is to employ instructional activities that provide students with an experiential basis to address better their own questions.

Pedagogical Resources for Exploring UIM

As I indicated, students from different disciplines in my philosophy of mind course question various aspects of UIM's connection of philosophy, science, and AI. In dealing with these questions, it should be admitted that the two paths to UIM, one mainly via Fodor and Dennett and the other via Sloman, are not entirely consistent. One point at which they do not cohere concerns the relationship between conceptual analysis and ordinary mental concepts. Sloman (1978) sees the explication of new possibilities resulting from the analysis of everyday mental terms, which in turn are tied to true background theories. In his later work, however, Sloman appears to back away from this position. With respect to mental concepts, there is no talk of ordinary language analysis leading to true background theories, and there is an openness to the possibility of scientific constructs reforming ordinary mental concepts. Folk-psychology concepts, it is admitted, may be contradictory and confused and serve only as "first approximations" (Wright, Sloman, and Beaudoin 1996, 107). Taking this point of view, which is tantamount to that of Fodor and Dennett, provides one means of supporting the coherence of UIM. Other points of difference arise, however, with respect to the function of explanation, prediction, and testability. Sloman de-emphasizes the role of explanation and prediction within science in favor of an interpretive role, one in which the discovery of possibilities is paramount and empirical testability is secondary.

Any philosophy of science that assumes that theories must be directly or easily testable is ill-conceived: it will fail for complex information processing systems, most of whose behavior is internal and unobservable. Moreover, even knowing how the system works may not provide a basis for predicting particular behaviors if the behavior depends not only on the design and current circumstances, but also on fine details of enduring changes produced by a long previous history. (Wright, Sloman, and Beaudoin 1996, 103)

This view of testability provides protection for very general possibilities that, not being well linked to brain neurophysiology or observable behavior, might not otherwise be scientifically acceptable. Perhaps this is a response to Popper's requirement of falsifiability, which would be unsatisfied in this context, but in any event it is misdirected. If one takes Fodor's inferred-entity approach to mental processes and Dennett's notion of simulations slowly progressing from abstractions to detailed instantiations, there is no need to minimize the role of testability. Initial testing involves conceptual analysis of the type referred to by Dennett (determining what kind of capabilities obtain from what designs, what interactions result with other components in a wider system, and so forth). Testability can thus involve conceptual coherence (internal behavior), and eventually empirical consequences (external behavior and instantiation). Taking a long-run view of theory development precludes the need to diminish the importance of testability. Perhaps those who have seen this the clearest are the cognitive-science pioneers themselves. The construction of programs like General Problem Solver was novel in its aim of providing a sufficient means to perform that task being explained. What was evident, however, was that this "sufficiency analysis" provided a first step toward a more detailed theory that would emerge amid the gross constraints of human information processing. "If an information processing system meeting these constraints can be devised that does the task, one then attempts to develop a revised system that has higher fidelity to specific data on human processing" (Newell and Simon 1972, 13). This sounds similar to Dennett's later statement on meeting growing constraints that provide greater psychological reality for continuously refined simulations, and it certainly reinforces a long-run view of theory development that values testability relative to stage of development.

Another fundamental point concerns the conditions under which a simulation of some cognitive function serves as an explanation of that function. Without a plausible link between explanations and simulations, AI remains weakly connected to both philosophy and cognitive science. Fodor supplies a link by rejecting major parts of the Ryle-Wittgenstein approach to linguistic meaning and its analysis. Sloman and his colleagues build upon this approach. Fodor's alternative is the inferred-entity model of mental concepts. Dennett, too, rejects this Wittgensteinian tradition, and his connection between simulations and explanations is less theoretical and more practical than Fodor's. Dennett offers a means by which simulations evolve into explanations, but this process offers no guarantee that the explanations arrived at are veridical, no matter how detailed the constraints. Fodor states that such explanations, like all explanations in science, serve at best as well-confirmed hypotheses. This view secures a connection between philosophy, science, and AI, but this approach is not available to Sloman. His program builds upon the Ryle-Wittgenstein view of mental concepts, and his de-emphasis of the explanatory-predictive role of science raises doubts about how simulations can be connected to explanations. Without this connection, the ability of AI to be closely related to science is in jeopardy. (Green [2000] provides an overview of the more recent status of the explanation-simulation quandary.)

It is difficult to overestimate the impact of accepting the inferred-entity approach to mental concepts. This view affects the aims and means of conceptual analysis and supports the view that the philosophy of mind has become a branch of the philosophy of science. As a consequence, many questions in the philosophy of mind, for example, concerning qualia, mind-body interaction, and so on, become versions of the question concerning the ontological status of hypothetical constructs. Even conflicts between realism and idealism, monism and dualism, and so forth are to be settled by whatever means ultimately settles the issue of the truth of scientific theories. Discussing this point in the classroom is important. Students can accept or reject these assertions, but they are the assertions that must be confronted if UIM is to be founded on the inferred-entity approach to mental concepts. This approach also leads to a view of conceptual analysis that does not fit well with that traditionally employed in philosophy. In particular, students should realize that mental concepts under ordinary language analysis often assume empirical conditions that are not readily observable. From the standpoint of a genuinely empirical inquiry these conditions should not be taken as given. As a simulation-mediated form of deducing the consequences of possibilities (hypothetical constructs), conceptual analysis differs from the form employed in ordinary language analyses. To succeed, science must be free to fashion technical concepts to explain and predict behavior within both experimental and natural settings. Conceptual analysis, whether carried out by philosophers or scientists, can provide those concepts.

These points may be contentious, but their explicit consideration in the classroom is essential. In particular, comparing two approaches to UIM serves to highlight the key issues, to put students in a position to accept or reject claims concerning these issues, and to address an array of roadblocks to understanding that arise from differences in disciplinary frameworks.

Innovations in Pedagogical Activities

Exploring UIM with differently trained groups of students requires attention to the questions raised by each group. Thus far, the emphasis has been on articulating certain contentious issues undermining the coherence of UIM. There is nothing new in the way of pedagogical technique in this. Another approach to matching instruction to student needs does, however, involve pedagogical innovation and the use of computers. This approach is to provide hands-on experience that will allow students to understand better and to address central issues and their own questions. In some cases, these experiences can be enhanced through the use of computers, particularly though interactive web sites. For example, in support of his own cognitive theory Dennett cites the Phi phenomenon, a simple visual illusion of motion. This phenomenon can be produced by flashing an image, such as a colored circle, at one location on a screen (point A) followed by another image flashed at a separate location (point B). Given appropriate timing, the circle appears to move from point A to point B. This experience occurs despite the fact that the circle is never displayed at any location between points A and B. In the classic Phi phenomenon, one is keenly aware both of the sense of motion and of the sense that the circle never appears in the interval it is crossing. In the related phenomenon of Beta apparent motion, one's sense of motion includes a glimpse of something crossing the interval (Palmer 1999). Student comprehension of this illusion can be facilitated by being granted control over the relevant timing variables. Figure 1 shows a simple graphic interface that grants this control in milliseconds. (The first of the two circular images has just been flashed above point A.)

Although research into apparent motion is carried out using a tachistoscope, microcomputers are sufficient for producing both Phi and Beta apparent motion. Having these experiences opens a path to engaging both the theoretical and the methodological issues surrounding Dennett's claims. Dennett (1991) criticizes two accounts of apparent motion (the Stalinesque and the Orwellian) while advancing his own account (Multiple Drafts). Figure 2 illustrates the Stalinesque account via a series of discrete, temporary events (T1 to T7), each lasting one hundred milliseconds or so. This account claims that there is a delay between the initial brain representation of the two flashing circles and the time when those circles are experienced in consciousness. Prior to the time delay, there is brain representation but no consciousness. After this delay, the area between the left and right circles contains a fleeting image that contributes to the experience of motion. This sequence of events reveals a kind of preconscious editing of conscious experience. Figure 3 illustrates the second account criticized by Dennett, the Orwellian account. This account claims that there is no delay between the brain's representation of the circles and the first conscious experience of them. Initially, the experience is of two circles separated in time and distance with nothing interposed between. No motion is sensed at this time. But immediately thereafter the experience is recorded in memory so as to include the fleeting interposition and the sense of motion that comes with it. So, immediately upon having the experience, one's identification and interpretation of it includes a sense of motion, and the immediate answer to the question "What was that?" is "a circle moving from one point to another." There is no memory of the initial experience that lacked a sense of motion.

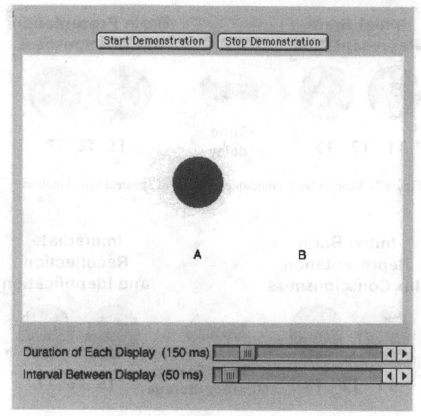

Figure 1. Interface for computer control of timing variables that regulate apparent visual motion.

Dennett believes that both of these accounts mistakenly rely on the "Cartesian Theater," which assumes that some area of the brain is responsible for producing and temporally organizing conscious experience. His alternative is to treat consciousness as a collection of parallel processes in which independent agents contribute to continually emerging drafts. Dennett asserts that we could never determine whether the experiential revisions postulated by the Stalinesque and Orwellian models actually occur. In fact, we can have no basis for accepting one model over the other. This claim underscores the connection of these accounts to issues of methodology. Flanagan (1992), for instance, sees this quandary as part of Dennett's attack on the purported role of qualitative concepts (qualia) in cognitive theories. In reductio fashion, reliance on qualia leads to unacceptable consequences, such as being unable empirically to distinguish

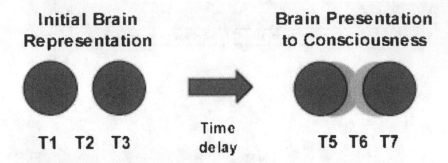

Figure 2. Time Delay (Stalinesque) account of apparent visual motion.

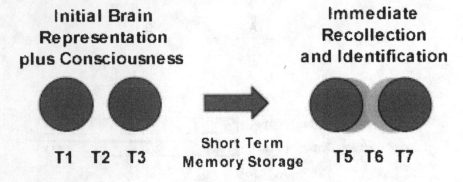

Figure 3. Memory Corruption (Orwellian) account of apparent visual motion.

between two conceptually distinct models. Flanagan maintains that Dennett is committed to the view that neurophysiological knowledge of how cognition works will never allow us reliably to identify cognitive events occurring on the order of milliseconds. Flanagan disputes this. In doing so he advocates the "natural method," which integrates phenomenological, behavioral, and neurophysiological approaches. This approach expects that the detailed neural workings of memory and conscious experience will one day be discovered, thereby making the Stalinesque and Orwellian accounts empirically distinguishable.

In addition to these considerations, other aspects of apparent motion also lead to questions concerning appropriate method. For example, a small percentage of students report not being able to see either Phi or Beta motion. Moreover, Dennett and other philosophers routinely cite a study of color changes accompanying the illusion of motion (Kolers and von Grunau 1976) in which the researchers use themselves (plus a trained student) as subjects for the experiment. Questions concerning individual variation and the appropriateness of scientists generating their own subjective reports hark back to challenges to introspectionism in nineteenth-century psychology. These issues plus Flanagan's discussion of methodology fit nicely within an interdisciplinary classroom. Overall, having direct experience of Phi and Beta apparent motion facilitates comprehension of these phenomena, and while not directly settling theoretical and methodological questions, encourages a motivated and thoughtful approach to the issues.

A similar benefit accrues to student interaction with programs that simulate or approximate cognitive abilities. With respect to ELIZA-like programs for natural language comprehension, students get a first-person understanding of how far conversation can and cannot be carried on using this syntactic technique. Interaction with other, more sophisticated communication programs is also helpful in this way. In addition, cognitive approaches to problem solving rely heavily on state-transition models along with diagrams of search spaces and descriptions of search techniques. When studying problem solving, computer programs can support an environment in which alternative states can be generated and searched, thus supplying a working definition of a problem as starting state, goal state, and transition rules. Standard examples, such as Cryptarithmetic, Missionaries and Cannibals, and Jealous Husbands, readily supply this foundation. Also, one issue in the machine-intelligence debate focuses on the power of heuristics and the level of performance attainable from simple search routines. Watching the Towers of Hanoi puzzle being quickly solved by a simple algorithm or competing against a computer program when playing Last One Loses, checkers, or chess is relevant to making judgments about how much intelligence should be expected from such approaches and to discussions of the distinction between formal and informal problems. Most important, these experiences provide a means by which general concepts and judgments can be instantiated in a direct fashion. Educational research supports the principles of interactive learning, and those principles are particularly relevant when learning about processes or techniques. Implementing the principles provides both cognitive and motivational benefits.

It has sometimes been remarked that with respect to cognitive simulation computers are in principle dispensable but in practice crucial. This may also be true of the pedagogical use of computers. In any case, computers offer opportunities that should not be neglected in the attempt to explain and assess UIM. That said, there exist useful and relevant pedagogical activities that are not technology based. Classroom exercises of all sorts can be employed in testing ideas about UIM. One can critically

explore, for example, the proposal that simulations can smoothly be transformed into explanations. In my class this exercise runs as follows.
Students are divided into three groups. Two of these groups read a description of an event to be explained. The description normally involves a crime
scene, such as a museum from which precious objects have been stolen
despite tight security. (These details can obviously be supplied as the
instructor sees fit.) One student group is also given three or four clues,
such as physical evidence at the crime scene. The task for these students is
to generate a hypothesis describing how the theft occurred in such a way
that all clues are accounted for. A second group of students is provided
with the same crime-scene description, minus the extra clues. The
students' task is not to explain how the crime was actually committed but
to formulate their own plan for successfully carrying off the theft. Then the
students in this group are sequentially given the clues and asked to revise
or elaborate their plan of attack in order to accommodate these new empirical conditions. Step by step, clues are added and plans reformulated.
Students in a third group play the role of critic and general evaluator. Their
job is to rank the accounts and plans with respect to general plausibility. To
do this, they generally request additional details and search for neglected
consequences and faulty inferences. Overall, this exercise provides
students with experiences that inform their discussion of the relation
between explanations and simulation.

If the tenets of UIM are considered to embody important disciplinary
aims and methods, these aims and methods should be taught in the philosophy of mind classroom. The question concerns how best to teach them.
This again depends upon local curricular factors, but to the degree that
students are expected genuinely to grasp the nature of these techniques,
learning should exceed mere reading about them. We would never expect
students to master logical analysis without engaging in numerous exercises, and we should not expect students fully to comprehend UIM by
simply reading about the accomplishments of those like Pollack (1998)
and Thagard (1998). When philosophical method becomes more empirical,
philosophical inquiry becomes more active and more process oriented.
Learning philosophy thereby involves doing and practicing, with the aims
and methods being taught. From an instructor's perspective, this may be
challenging, but a class comprising students from different backgrounds
proves to be a useful resource. One possible exercise would be to form
student teams, ideally combining at least philosophy, psychology, biology,
and computer-science majors. Each team would be given the task of
designing some cognitive component, first at an abstract level and then in
as much detail as possible. The details of the component's performance
might be designated by observing the behavior of a simple organism or
some aspect of its interaction with the environment.

The empirical orientation of contemporary philosophy of mind has
attracted the attention of faculty in other disciplines who see value in

having their students enroll in a relevant philosophy class. Beyond this, there are disciplines whose own course material has philosophical connections. In these cases, philosophers can make contributions to a variety of topics, and perhaps the best examples of these are courses in cognitive science. Cognitive-science curricula are thriving in many academic quarters. Philosophical emphasis on empirical theories of mind has established philosophy's relevance to the cognitive-science curriculum. Philosophical perspectives are sought after in what is universally recognized as an interdisciplinary enterprise.

Some of these courses have associated web sites for interactive exploration of various topics. Examples of such web sites are becoming easier to find. At my own university the Introduction to Cognitive Science course involves faculty from psychology, philosophy, computer science, and linguistics. The course is supported by an interactive web site that contains modules on a variety of topics, including those such as consciousness, machine intelligence, problem solving, and natural language understanding, which have explicit philosophical connections (www.uncc.edu/mjcroy/cogsci.html).[2] In such cases, philosophical contributions and perspectives can be made explicit. Another pedagogically oriented example of philosophy in the cognitive-science curriculum can be seen in the Mind Project (previously ProtoThinker) maintained by John Barker and David Anderson at Illinois State University. This site also provides materials and activities that are useful in teaching philosophy of mind proper (www.mind.ilstu.edu). Another interesting site from a pedagogical perspective is the one maintained by Saul Traiger, a philosopher who directs the cognitive-science program at Occidental University. Here philosophical influence can be found on a variety of courses, particularly those with epistemological connections (www.traiger.oxy.edu/index.html).

Conclusion

To the extent that cognitive theory increasingly becomes part of the philosophical investigation of mind, it should be expected that philosophy of mind courses will increasingly become interdisciplinary endeavors that draw students from different disciplines. When these students come to the philosophy of mind class, confronting methodological issues becomes crucial. Perhaps the best way for students to appreciate what changes in method mean for philosophy and for its relation to other disciplines is to supplement reading with a more active, cooperative, engagement of the issues. Taking advantage of interactive web sites is one means of achieving

[2] Construction of this web site was funded by the National Science Foundation (9950736). The project team included members from four departments: Computer Science (Mirsad Hadzikadic, Jan Zytkow), English (Boyd Davis, Ralf Theide), Philosophy (Marvin Croy), and Psychology (Sarah Breedin, Paul Foos, Jane Gualtny, Paula Goolkasian, Lori Van Wallendael). Adam Garside programmed the graphic interface shown in figure 1.

this aim. Devising in-class activities that exploit diverse student talents provides another alternative. Engaging students in the actual process of designing and carrying out simulations offers still another helpful activity. If cognitive simulation is indeed a form of conceptual analysis and an indispensable part of the pursuit of empirically and conceptually defensible theories, such pedagogical activities are reasonable and perhaps essential. In recent decades, computers have contributed to an empirical orientation in the philosophy of mind both in terms of content and method. In the decades to come, computers may play an even more significant role in passing this orientation on to the next generation of philosophers.

References

Chihara, Charles, and Jerry Fodor. 1965. "Operationalism and Ordinary Language: A Critique of Wittgenstein." *American Philosophical Quarterly* 2: 281–95.

Dennett, Daniel. 1978a. *Brainstorms*. Montgomery, Vt.: Bradford Books.

―――. 1978b. "Current Issues in the Philosophy of Mind." *American Philosophical Quarterly* 15: 249–61.

―――. 1991. *Consciousness Explained*. New York: Little, Brown.

Flanagan, Owen. 1992. *Consciousness Reconsidered*. Cambridge, Mass.: MIT Press.

Fodor, Jerry. 1968. *Psychological Explanation*. New York: Random House.

Green, Christopher. 2000. "Is AI the Right Method for Cognitive Science?" *Psycoloquy* 11: 61. Also available at www.yorku.ca/christo/papers/pubs.htm.

Kolers, Paul, and Michael von Grunau. 1976. "Shape and Color in Apparent Motion." *Vision Research* 16: 329–35.

Newell, Allen, and Herbert Simon. 1972. *Human Problem Solving*. Englewood Cliffs, N.J.: Prentice-Hall.

Palmer, Stephen. 1999. *Vision Science: Photons to Phenomenology*. Cambridge, Mass.: MIT Press.

Pollack, John. 1998. "Procedural Epistemology." In *The Digital Phoenix: How Computers Are Changing Philosophy*, edited by Terrell Ward Bynum and James H. Moor, 17–36. Oxford: Blackwell.

Ringle, Martin. 1979. *Philosophical Perspectives in Artificial Intelligence*. Atlantic Highlands, N.J.: Humanities Press.

Ryle, Gilbert. 1949. *The Concept of Mind*. London: Hutchinson.

Sayre, Kenneth. 1979. "The Simulation of Epistemic Acts." In *Philosophical Perspectives in Artificial Intelligence,* edited by Martin Ringle, 17–36. Atlantic Highlands, N.J.: Humanities Press.

Sloman, Aaron. 1978. *The Computer Revolution in Philosophy: Philosophy, Science, and Models of Mind*. Atlantic Highlands, N.J.: Humanities Press.

Thagard, Paul. 1998. "Computation and the Philosophy of Science." In *The Digital Phoenix: How Computers Are Changing Philosophy*, edited by Terrell Ward Bynum and James H. Moor, 48–61. Oxford: Blackwell.

Todd, William. 1977. "The Use of Simulations in Analytic Philosophy." *Metaphilosophy* 8, no. 4: 272–97.

White, Alan. 1967. *The Philosophy of Mind*. New York: Random House.

Wright, Ian, Aaron Sloman, and Luc Beaudoin. 1996. "Towards a Design Based Analysis of Emotional Episodes." *Philosophy, Psychiatry, Psychology* 3: 101–26.

5

PHENOMENOLOGY AND ARTIFICIAL INTELLIGENCE

ANTHONY F. BEAVERS

Phenomenology as a term in philosophy and psychology is somewhat ambiguous. In a strict sense, it names a method for analyzing consciousness that was first developed by Edmund Husserl. In a broader and more relaxed sense, it also refers to general first-person descriptions of human experience. In this second sense, a "phenomenology" would undertake to describe an experience like pain perhaps as the throbbing, dull, or sharp sensation that distracts me from my daily tasks, or depression as a listless indifference to anything that could make a difference. Phenomenological descriptions, or "phenomenologies" taken in this relaxed sense, are often contrasted with scientific explanations of the same phenomena. A scientific account of pain, for instance, speaks of C-fiber firings and neurological activity, not significantly of sharp or dull sensations.

The applicability of phenomenology for artificial-intelligence and cognitive-science research is seriously limited if we take the term in this second sense. The reason is simple enough; the genuine causes of conscious phenomena and other internal states do not themselves appear in first-person experience. The apparent seeming that characterizes internal experience does not provide access to the genuine causes that operate "beneath." Thus, what it "feels" like to see blue or what we think we are doing when we entertain a belief simply does not reach to the level of scientific explanation. If our goal is to learn about how intelligence and other cognitive activities actually operate, phenomenologies of experience can at best be starting points or guiding points, useful to the extent that they better acquaint us with the phenomena we hope to explain or the effects for which we seek the causes.

This view is true enough if we take phenomenology in the relaxed sense described above. But it does not seem to hold for phenomenology in the narrow sense, that is, as a method for describing consciousness, first formalized and advanced by Husserl. The purpose of this essay is to show why this is so.

I wish to thank Larry Colter, Julia Galbus, and Peter Suber for their comments on earlier drafts of this essay.

Microworlds and Cognitive Modeling

A central theme of classical phenomenology is something called "world constitution," the process whereby consciousness builds its conception(s) of, or "makes known," a world. This is a rich and varied aspect of phenomenology, and we cannot expect to do justice to it here. Much of what follows, however, will relate to this notion, primarily because of what it means for the "microworlds" approach to artificial intelligence that has been a part of the discipline since its inception. This approach was analyzed and criticized as early as 1979 by Hubert Dreyfus in his essay "From Micro-Worlds to Knowledge Representation: AI at an Impasse," and it remains a relevant aspect of AI research even in as recent a text as Luciano Floridi's 1999 book *Philosophy and Computing: An Introduction.*

The purpose of a microworld in AI is to build a closed domain of virtual objects, properties, and relations small enough to be sufficiently mapped by a computer. The fact that the domain is closed helps the computer to "determine the meaning" of commands that might otherwise be ambiguous, so that it can respond appropriately in its closed environment. In addition, because world parameters are limited and only a small number of objects, properties, and relations obtain in this environment, the programmers' task is also greatly simplified. They need not concern themselves with every possible contingency that could obtain in the "real" world, only those that do obtain in the microworld.

Both Dreyfus and Floridi raise the objection that because computers, or computer programs, are locked in microworlds and human beings are not, AI research cannot approximate human intelligence, which is open ended and able to deal with a broad range of contingencies. The two accounts differ in emphasis, however, the first taking an epistemic approach and the second an ontological. Because the ontological reading is more useful for my purposes, I shall base my comments on Floridi's critique. I shall address Dreyfus's view in a later essay.

Floridi writes:

> The specific construction of a microworld "within" a computerized system represents a combination of ontological commitments that programmers are both implicitly ready to assume when designing the system and willing to allow the system to adopt. This tight coupling with the environment (immanency) is a feature of animal and artificial intelligence, its strength and its dramatic limit. On the contrary, what makes sophisticated forms of human intelligence peculiarly human is the equilibrium they show between creative responsiveness to the environment and reflective detachment from it (transcendency). This is why animals and computers cannot laugh, cry, recount stories, lie or deceive, as Wittgenstein reminds us, whereas human beings also have the ability to behave appropriately in the face of an open-ended range of contingencies and make the relevant adjustments in their interactions. A computer is always immanently trapped within a microworld. (1999, 146–47)

Apparently, according to Floridi, computers – unlike humans – live in a world in which they are bound by some type of immediate ontological necessity. We human beings, on the other hand, exhibit a freedom that liberates us from (some of) the ontological ties to our environment. Later in the book, the importance of this ontological difference becomes apparent with regard to AI research. Floridi writes: "The more compatible an agent and its environment become, the more likely it is that the former will be able to perform its task efficiently. The wheel is a good solution to moving only in an environment that includes good roads. Let us define as 'ontological enveloping'[1] the process of adapting the environment to the agent in order to enhance the latter's capacities of interaction" (1999, 214). Here Floridi correctly implies that the degree of ontological enveloping is directly proportional to the success of an artificially intelligent agent. He cites web-bots as good candidates for success in this area, because they are digital agents living in a digital world. Within this context, Floridi suggests (1) that success for artificial agents requires ontological enveloping, whereas our "transcendency" makes success for human agents possible along different lines, and (2) that the principles that govern artificial intelligence should, therefore, be different from those that govern human intelligence. In fact, this discussion appears as part of an extended argument for preferring "light," or non-mimetic, artificial intelligence (LAI) to more traditional or mimetic attempts at AI (GOFAI).[2]

One problem with such a view is that it renders useless (for AI

[1] The meaning of the word *enveloping* is difficult to discern in this context, though the definition of the process seems clear. It is "the process of adapting the environment to the agent in order to enhance the latter's capacities of interactions." "Ontological accommodation" might be a better term, but I shall keep with Floridi's usage throughout. Readers will need to remember that it is the process as defined that I mean to indicate by the use of Floridi's terminology and not be too concerned with making sense of the concept of enveloping.

[2] "Mimetic" approaches to AI are those that try to build artificial agents that do things the way human beings do. "Non-mimetic" approaches strive for the same result by attempting to emulate thinking along functionalist lines. Floridi's critique is based on the early analysis of Turing. "Since GOFAI could start from an ideal prototype, i.e., a Universal Turing machine, the mimetic approach was also sustained by a reinterpretation of what human intelligence could be. Thus, in Turing's paper we read not only that (1) digital computers must simulate human agents, but also that (2) they can do so because the latter are, after all, only complex processors (in Turing's sense of being UTMs)" (Floridi 1999, 149). Evidence that Turing makes a mimetic mistake is apparent in his admittedly bizarre attempt to estimate the binary capacity of a human brain. (See the closing section of *Computing Machinery and Intelligence*.) Still, though Turing thought of the human being as a UTM, thereby committing a mimetic fallacy, we need not conclude that all mimetic approaches are doomed to fail. It is conceivable and may even be likely that connectionist networks will one day model human cognition at a reasonable level of complexity. Details aside, they do seem to mimic brains, at least minimally. Because these networks are computationally equivalent to UTMs, it is premature to give up on the idea that the brain is a computer.

purposes) philosophical systems based on doctrines of transcendental world constitution, like those advanced by Kant and Husserl. What is the *Critique of Pure Reason*, if not an attempt to isolate the necessary rules that must be enacted on sensibility in order for there to be experience of objects? On a broad reading of Kant's text, scientific inquiry is possible precisely because the rules used for "ontological enveloping" mirror the fundamental laws of science, allowing for a goodness of fit between agent and world analogous to the wheel example mentioned above. Floridi's view also renders insignificant (for AI purposes) the phenomenological work of Husserl, which is significantly bound up with a similar doctrine of world constitution in an attempt to articulate the cognitive architecture that makes possible a world of experienced (and experience-able) objects.

Of course, rendering a philosophical system useless for specific purposes is not a mistake if good reason can be found for doing so. Here the reason seems to be based on the apparently obvious fact that human experience is open ended and therefore not reducible to a microworld. It is detachment from the world (or transcendency) and not immanency that makes the human condition unique. Floridi writes:

> We must not forget that only under specially regimented conditions can a collection of detected relations of difference [binary relations, for instance] concerning some empirical aspect of reality replace direct experiential knowledge of it. Computers may never fail to read a barcode correctly, but cannot explain the difference between a painting by Monet and one by Pissarro. More generally, mimetic approaches to AI are not viable because knowledge, experience, bodily involvement and interaction with the context all have a cumulative and irreversible nature. (1999, 215–16)

A decent pattern-matching procedure to articulate the difference between paintings certainly seems possible—after all, computers are capable of facial recognition—and it is not readily apparent why open-endedness should immediately lead to the conclusion that human intellectual success entails our not living in microworlds. Coping with contingencies could involve procedures whereby a new encounter or experience is mapped into an existing representational structure as the human agent builds or modifies its microworld. Such procedures have been outlined by several representatives of the post-Kantian tradition, though perhaps not in a language recognizable to AI and cognitive-science researchers. At very least, microworld philosophers, like Kant and Husserl and others who hold to doctrines of transcendental world constitution, take the world to be ontologically enveloped even in the face of open-ended human contingencies, if I can translate their view into Floridi's language.

If I am correct in this reading of the phenomenological tradition, even before Husserl, then it is premature to conclude that artificial intelligence cannot approach genuine human intelligence, *even if machines are trapped in microworlds*. Before reaching this conclusion, a thorough and in-depth look at the utility of these systems for AI purposes is necessary. Of course, we cannot undertake this task here; but we can hope to show the legitimacy and importance of such an initiative.

Much of Floridi's rationale for refusing the microworld approach to human cognition seems to hang on something he calls the "sigma fallacy." He states:

> [I]f an intelligent task can be successfully performed only on the basis of knowledge, experience, bodily movement, interaction with the environment and social relations, no alternative non-mimetic approach is available, and any strong AI project is doomed to fail. To think otherwise, to forget about the non-mimetic and constructive requirements constraining the success of a computable approach, is to commit what we may call the Σ fallacy and to believe that, since knowledge of a physical object, for example, may in general be described as arising out of a finite series of perceptual experiences of that object, then the former is just a short hand notation for the latter and can be constructed extensionally and piecemeal, as a summation. (1999, 216)

It is difficult to tell from the context how extensively Floridi means these words, and much hangs on the word *just* in the last part of the sentence: "the former is *just* a short hand notation for the latter." Take out the word *just* and replace it with "more or less," and we find ourselves implicitly accepting some form of transcendental constitution. Such theories do suggest that data from perceptual experience are picked up by consciousness and put through a series of operations to construct a "world of objects" that serves to explain what appears to us in experience. For phenomenologists, the phenomenal world *is* just such a summation, a spatial and temporal objectification of the flux of sense data into a stable and knowable world according to a set of processes or procedures. Of course, "summation" must be taken in a loose sense here, for in the process of world constitution, consciousness may ignore some sense data, retrieve others from memory, confuse sensations with perceptions, and so on. In fact, some theories, such as Husserl's, for instance, are quite intricate in their descriptions, making "result," "function," or even "computation" better words than "summation."

Do such views commit the sigma fallacy? It is difficult to say. Even so, the basic theoretical disposition among them remains the same; there *are* discoverable rules that govern the process whereby consciousness transforms the flux of sense data into a world of possible objects of experience. The import of this observation for AI should be clear; where we find rules, we find the possibility of algorithms. Clearly, a formalized set of such rules would be relevant to research in artificial intelligence and cognitive science.[3]

---

[3] I am not claiming that Kant or Husserl, or anyone else for that matter, has furnished a complete or even minimally adequate set of rules for world constitution. My point is only that AI and cognitive-science researchers should want these efforts to succeed. In addition,

The Phenomenological Reduction as Preliminary to a Science of Cognition

The key to understanding the significance of phenomenology for our purposes lies in its broad outlines, at its very beginning in the phenomenological reduction, or *epoché,* as it is sometimes called. This tactic permits us to make a useful distinction between two attitudes that we can take toward the "world" and, consequently, the role it is allowed to play in explaining the various phenomena that appear within experience. It is enacted by Husserl *"to make 'pure' consciousness, and subsequently the whole phenomenological region, accessible to us"* (1982, 33). But, as we shall see, its significance for us lies in the way it allows us to draw a further distinction between the world in which human intelligence operates and the intellectual architecture that ontologically envelops such a world.

To undertake the phenomenological reduction, we need only consider our experiences *as if* they had no existential counterparts in a world outside consciousness. Existence, for the moment, is reduced to its pure appearances within consciousness, to pure phenomena, in the strict, etymological sense of the term. The rationale for this move is purely methodological. It "shall serve us only as a *methodic expedient* for picking out certain points which . . . can be brought to light and made evident by means of it" (Husserl 1982, 31). It is meant to suspend the positing of a *"factually existent actuality"* (Husserl 1982, 30) that characterizes our natural way of taking the world. As such, it suspends our deeply rooted judgment that the world is "there" as an ontic and extra-mental unity. This is important because "the single facts, the facticity of the natural world taken universally, disappear from our theoretical regard" (Husserl 1982, 34).

What remains after the reduction is the terrain proper for phenomenological investigation, consisting of pure mental processes, pure consciousness, the pure correlates of consciousness, and the pure ego (Husserl 1982, 33). As such, the reduction isolates the cognitive components of our general experience of the natural world and makes them available for description.

How does it do so? A brief phenomenological experiment will help to clarify the process. Sense data provide us with a multiplicity of appearances such that at T_1 I may have a visual perception of a computer screen before me, turn away, and at T_2 have another visual perception of a computer screen. I can then repeat this experience at T_3 and T_4. On the level of sense, four separate experiences occur at four separate times. Yet

I agree that traditional attempts at understanding world constitution are too representational, a charge that has clearly been continually waged against Husserl by thinkers like Martin Heidegger and Emmanuel Levinas. But this, by itself, does not mean that the representational account is rendered useless. It means only that it must be situated. Even if embodied action is a necessary condition for human or mimetic representation, a cognitive layer of data processing seems nonetheless to be operative.

I do not conclude that there are four separate computer screens, one for each sensory experience. Why not? It would seem that a strict empirical counting off of four separate experiences should lead us to four separate objects.

From the standpoint of the natural attitude, that is, operating from a perspective before the reduction, we can appeal to "laws of nature" and general ontological commitments to reach the conclusion that because each of our perceptions was caused by an extra-mental computer screen that is actually "out there," there is only the one. This conclusion may well be correct, but the point here is not to establish what really is the case. Instead, we are trying to learn what we can about our cognitive architecture.

To do so, we switch to the phenomenological viewpoint and explain how four separate sensory experiences are transformed into belief in a single, unified computer screen out there able to cause all four of them. Because we cannot leap over sensibility to the thing-in-itself and our foundational evidence is fourfold, we must turn inward to some process of cognition that simplifies or synthesizes the multiplicity of appearances into a single object that we can posit behind the appearances to explain their regularity.

The critical point here is to understand that our judgment (that a permanent object exists behind sensibility) is structured according to an implicit ontology that is posited by our scientific conception of nature. This ontological commitment becomes visible as such when we consider the situation from the phenomenological attitude. Here the experience itself is our starting point, and because we have no access to the extra-mental, save through sensibility, the permanence of an object is something that we determine internally and then assign to the objects in the world. In other words, we know it insofar as it is derived from our cognitive processes, not because we find it in the world.

Of course, actual phenomenological description is much more subtle and detailed, but this crude example should serve to illustrate the cognitive significance of the method, namely, that the objects (and, to generalize, the properties and relations between them) that were previously taken to belong to the ontology of the natural world now appear as cognitive structures that result from mental processes directed at sensibility. The previous objects of nature turn out to be "naturalized objects," that is, ideal objects that are constituted on the basis of appearances in concrete experience and projected outward into the natural world.

Husserl's first examples understand this process of naturalization in reference to material objects. The perceptions of a material object arise one at a time in a succession of appearances or perspectives that leads to the conclusion that "behind" these appearances there exists a thing-in-itself that causes them. Naturalism treats these things-in-themselves as if they were absolute objects, failing to treat them in relation to the perceptions from which they were abstracted:

By interpreting the ideal world which is discovered by science on the basis of the changing and elusive world of perception as absolute being, of which the perceptible world would be only a subjective appearance, naturalism betrays the internal meaning of perceptual experience. Physical nature has meaning only with respect to an existence which is revealed through the relativity of *Abschattungen* [perspectives]—and this is the *sui generis* mode of existing of material reality. (Levinas 1973, 10)

The process as a whole seems to be marked by a manifest circularity. Our ideas of objects are caused by sense data as processed by cognition. These objects are also taken to be out there in the world functioning as causes of sense data. These objects then both cause and are caused by sense data. But we can make reasonable sense out of this circularity by appeal to the notion of the two attitudes discussed above. In the natural attitude, the world of natural objects *is* out there functioning as the cause of our sense data. In the phenomenological attitude, the picture is inverted and objects are revealed as information structures that unite a multiplicity of appearances according to various mental acts. From this attitude, science is then seen to be operating within cognitive structures that are posited as belonging to the world. These structures provide the very ontology of the natural world, or, in other words, the natural world of science is ontologically enveloped (to use Floridi's term) in advance by consciousness as a precondition for being able to frame cognitive claims about it.

This ontological enveloping pertains not only to material objects but to ideal objects as well. "Besides consciousness [and material objects], naturalism is also obliged to naturalize everything which is either ideal or general—numbers, geometrical essences, etc.—if it wants to attribute to them any reality at all" (Levinas 1973, 14). Levinas goes on to say,

[T]he specific being of nature imposes the search, in the midst of a multiple and changing reality, for a causality which is *behind* it. One must start from what is immediately given and go back to that reality which accounts for what is given. The movement of science is not so much the passage from the particular to the general as it is the passage from the concrete sensible to the hypothetical superstructure which claims to realize what is intimated in the subjective phenomena. In other words: *the essential movement of a truth-oriented thought consists in the construction of a supremely real world on the basis of the concrete world in which we live.* (Levinas 1973, 15–16)

We see then that one consequence of the viewpoint of natural science is that it disguises the informational content that is derived from these data and the conceptual architecture that processes them as the ontology of the external world. The purpose of the reduction is to make what is hidden in the natural viewpoint apparent. "As long as the possibility of the phenomenological attitude had not been recognized, and the method for bringing about an originary seizing upon the objectivities that arise within that attitude had not been developed, the phenomenological world had to remain unknown, indeed, hardly even suspected" (Husserl 1982, 33).[4]

Once we understand that *all of our evidence* about the external world comes through sense data that *we* process into a coherent picture of reality, it should be clear that this coherent picture is the informational output of processes that transform the input of sense data into a world representation. This phenomenological representation *is* the extra-mental world *as we take it to be*.

We see then that the term *extra-mental* has two senses, one of which refers to what is genuinely out there in some Kantian world of things-in-themselves, and the other of which is within consciousness as a "transcendence in immanence,"[5] a projection of phenomenological objectivities into the world of natural science. Immediately, it will seem that phenomenology as articulated here is a type of idealism, so much so that thinkers like Paul Churchland, for instance, define phenomenology and idealism as inherently interrelated. (See Churchland 1988, 83–87). But we need not go that far.[6] For it could well be the case that the phenomenological world, the transcendence in immanence constituted out of sense data by mental processes, maps adequately onto the world of things-in-themselves. Just maybe we get things right. The decision of whether or not it does map in this way and the extent to which it does (and whether and how this is at all knowable) will determine whether we are on realist or idealist grounds. But we need not make any dogmatic metaphysical and epistemic commitments here, even while suggesting that what is genuinely out there can affect us only through our senses. We can still use the method of phenomenology for making explicit and describing cognition.

If human beings did not have sight but did have echolocation and a keen pheromone sense, our world as a transcendence in immanence, our

---

[4] Numbers in the references to Husserl refer to sections in his *Ideas* and not to the actual page numbers.

[5] Husserl uses this expression to speak of a world that is genuinely immanent, that is, built up within consciousness, but projected outward in such a way that it looks extra-mental. For a sustained treatment of "transcendence in immanence," see the fifth of Husserl's *Cartesian Meditations*.

[6] The issue of the relationship between idealism and phenomenology has a long and complicated history of its own. For a sustained commentary, see the lengthy treatment by Herman Philipse in *The Cambridge Companion to Husserl* (Philipse 1995, 239–322). There is no doubt that Husserl is a transcendental idealist, and Philipse is correct to note that "Husserlian phenomenology without transcendental idealism is nonsensical. If one wants to be a phenomenologist without being a transcendental idealist, one should make clear what one's non-Husserlian conception of phenomenology amounts to, and which problems it is meant to solve" (Philipse 1995, 277). I do believe that Husserl's phenomenology needs to be reworked in light of the problem of idealism, but it is unfortunate that the possibility of idealism is immediately taken to mean that phenomenology as such is bankrupt as a tool for learning about cognitive processes. We can use it as a method, temporarily accepting its ontological commitments, without having to posit these commitments as binding outside the practice of the method.

"vision" of reality, would be quite different from what it currently is, even if the genuine extra-mental world and our mental procedures for processing sense data were to remain the same. The problem would still be to explain how this vision is acquired, how the "world for us" is wrenched out of the data stream. This issue is independent of the actual states of affairs outside cognition, whatever they may turn out to be.

The critical element here is that because our vision of reality must go back to sense data, and there is a difference between our knowledge about this world and what is available in sense data, cognitive processes are somewhere involved. Phenomenology is useful to the extent that it makes these cognitive processes available for analysis and description. But if this is the case, then the world in which human beings operate is already a microworld posited as, and taken to be, the genuine extra-mental world, even though it really is an ontological model of it. In other words, the extra-mental world *as we take it to be* is already ontologically enveloped.

The importance of the phenomenological method is not only that it makes this ontological enveloping explicit but also that it goes on to describe the various mental acts that are required to bring it about. This is not the place to engage in such a description. The point is only that phenomenology is already a science of human cognition, engaged in articulating the acts of consciousness that ontologically envelop the world of cognition and the ontological structures that emerge as part of this enveloping.

Conclusion

If I am correct in this assessment of the phenomenological enterprise, its applicability for artificial-intelligence and cognitive-science research should be clear. By articulating the mental processes involved in getting from sense data to a knowable world, it is engaged in understanding the processes and procedures involved in human cognition. While such an understanding does not explain cognition or how the brain works, it does help us understand which processes are instantiated in the brain. Such an understanding certainly is an aid to cognitive science. Furthermore, the same understanding can guide our efforts to duplicate intelligence in machines. After all, we need to know what we want a machine to do, before we can build one to do it, and it is insufficient to presuppose a set of mental acts without careful attention to their precise function within our cognitive initiatives.

Having said this, I mean in no way to suggest that we can simply apply Husserl to artificial intelligence. No doubt much of what he has said is relevant to our purposes, but it would take considerable effort to get what we need out of his dense and turgid prose. A shorter, and perhaps more useful, approach would be to borrow his method in the climate of current cognitive science and rework phenomenology, perhaps along realist lines, with different goals in mind.

Finally, I must address the issue of why, if I am correct about this application of phenomenology, such a view is not widely held. I can only speculate. A possible reason may be that the tendency of phenomenologists to write in the tradition of German philosophy, indeed German Idealism, confuses the issues or, at least, hides their significance for cognitive science. The language often suggests spiritual powers belonging to a disembodied psyche at the center of an ideal world—an ego playing God. This usage lends itself to idealistic interpretations of phenomenology that make metaphysical commitments instead of methodological ones. It does not help that Husserl himself may be guilty of such a charge. Realist phenomenologies friendly to materialism and recent work in cognitive technology are possible, even while moving in the wake of the phenomenological reduction. But this is not easily seen.

Thinkers like Hubert Dreyfus have used phenomenology negatively to argue that traditional approaches to artificial intelligence are doomed to fail (Dreyfus 1992). But understanding phenomenology as an aspect of cognitive science puts it to constructive use. Metaphysical commitments aside, phenomenologists like Husserl and Heidegger can be read as describing (in detail) the informational processes and procedures by which cognition operates. Although these descriptions may not be immediately in line with Newell and Simon's Symbol System Hypothesis or binary computation, it is a mistake to think that they cannot be translated into some form that can be modeled by intelligent technology. We should be reading these theories positively to see how AI can be possible. At very least, a formalized phenomenology of cognition following the appropriate method of description should prove worthwhile for artificial intelligence and cognitive science. It may well provide us with what we need to duplicate human intelligence in machines, even if it can never reach beneath appearances to understand the architecture and function of the human brain.

References

Churchland, Paul. 1988. *Matter and Consciousness*. Revised edition. Cambridge, Mass.: MIT Press.

Dennett, Daniel C. 1984. "Cognitive Wheels: The Frame Problem of AI." In *Minds, Machines and Evolution: Philosophical Studies*, edited by Christopher Hookway, 129–51. New York: Cambridge University Press.

Dreyfus, Hubert L. 1992. *What Computers Still Can't Do: A Critique of Artificial Reason*. Cambridge, Mass.: MIT Press.

———. 1997. "From Micro-Worlds to Knowledge Representation: AI at an Impasse." In *Mind Design II: Philosophy, Psychology, Artificial Intelligence*, edited by John Haugeland, 143–82. Cambridge, Mass: MIT Press.

Floridi, Luciano. 1999. *Philosophy and Computing: An Introduction*. New York: Routledge.

Heidegger, Martin. 1996. *Being and Time*, translated by Joan Stambaugh. Albany, N.Y.: SUNY Press.

Husserl, Edmund. 1960. *Cartesian Meditations: An Introduction to Phenomenology*, translated by Dorion Cairns. The Hague: Nijhoff.

———. 1982. *Ideas Pertaining to a Pure Phenomenology and to a Phenomenological Philosophy. First Book. General Introduction to a Pure Phenomenology*, translated by F. Kersten. The Hague: Nijhoff.

Kant, Immanuel. 1929. *Critique of Pure Reason*, translated by Norman Kemp Smith. New York: St. Martin's Press.

Levinas, Emmanuel. 1973. *The Theory of Intuition in Husserl's Phenomenology*, translated by André Orianne. Evanston, Il.: Northwestern University Press.

Newell, Allen, and Herbert Simon. 1990. "Computer Science as Empirical Inquiry: Symbols and Search." In *Mind Design II: Philosophy, Psychology, Artificial Intelligence*, edited by John Haugeland, 81–110. Cambridge, Mass.: MIT Press.

Philipse, Herman. 1995. "Transcendental Idealism." In *The Cambridge Companion to Husserl*, edited by Barry Smith and David Woodruff Smith, 239–322. New York: Cambridge University Press.

Turing, Alan. 1997. "Computing Machinery and Intelligence." In *Mind Design II: Philosophy, Psychology, Artificial Intelligence*, edited by John Haugeland, 29–56. Cambridge, Mass.: MIT Press.

6

ADAPTABLE ROBOTS

GENE KORIENEK and WILLIAM UZGALIS

Complex Environments

In this essay we consider some of the characteristics of adaptive biological systems and how these might work as models in designing a robot intended for the exploration of complex environments. This in turn helps make clear the connections among emergent properties, collective decisions, and adaptability. Trying to design a robot that has such properties forces one to think hard about the nature of those properties. Here we have an intersection between philosophy and computing.

Adaptability seems to be a fundamental characteristic of living organisms. One definition characterizes life as "subtle adaptation" – "the capacity to respond appropriately, in an indefinite variety of ways, to an unpredictable (from the perspective of the organism) variety of contingencies" (Bedau in Boden 1996). Adaptation is a basic and crucial strategy for survival in a changing world. Given this, it is worth noting that adaptation, at least on the level of the species, does not seem to require much subtlety or intelligence. Why is this? If an organism happens to have the right characteristic to operate in an environment, it survives. So, roughly speaking, the part of the species that has the favored characteristic tends to survive, the parts without it do not. So, the species changes. It adapts. But neither consciousness, nor a sense of making a choice, nor even having values seems to be required. There need be no desire to survive for adaptation to occur. Very likely it is for these kinds of reasons that computer models of evolution are so successful. Of course, if individuals are adaptive this would likely make the species more adaptive. What, then, is required for an individual (as opposed to a species) to be adaptive?

If we are interested in the survival of a single organism, the way adaptation works on the species level may not seem relevant. Adaptation on the level of the individual is not only a strategy for survival but is also conceptually connected with at least one other important characteristic of biological agents. An adaptive agent must be an autonomous, or self-directed, agent. If a robot is not adaptive, then any form of new behavior must come as a result of commands sent to it by radio signal from its controllers or be pre-programmed into its central processor. Surely it would be better for a robot to be able to change its behavior on its own to deal rapidly with new and unexpected information about the environment. Autonomy, then, seems to add a new dimension to individual adaptability not necessarily present on the level of species adaptability.

Another important point is that autonomy and adaptability come in degrees. Human beings are very likely autonomous in a way and to a degree different from an amoeba or a rodent. So, the less adaptive an organism or robot is, the less autonomous it is going to be. The more adaptive, the more autonomous. So, we are looking for those qualities that make one organism more adaptive than another. These degrees of autonomy and adaptability may be difficult to distinguish, but they are surely important. If it were necessary to produce the kind of powers that allow humans to adapt, powers associated with the central nervous system, it might require more of us as designers of machines than we can manage. We prefer to look to less intelligent systems. In the flocking of birds and the complex organization of insects one may find clues to a more modest yet still subtle adaptability. In human beings, we move from the central to the peripheral nervous system in search of such subtle adaptive mechanisms. Such mechanisms are so primitive they require neither memory nor internal representations of the world. One reason for pursuing this strategy comes from consideration of contemporary robotic design.

Conventionally engineered systems, both hardware and software, typically include a complete "specification" of the system as a part of their design. Any system behavior outside the specification is often considered an error. Such "specified" systems exhibit low levels of adaptability and are typically not autonomous. The issues plaguing robotic design for robots that need to be adaptive and autonomous lie in the mismatch between pre-programmed tasks and the kind of flexible behaviors necessary to manage behavior within the characteristics of a changing, and often unknown, environment. Thus, the kind of design that is being employed is more appropriate to industrial robotics, where the environment that the robot is going to operate in is completely understood and the actions it is going to take in the environment to achieve its goals are also completely understood. The problem is that this kind of design does not allow for encountering unknown obstacles and doing something different to get around them.

In *The Science of the Artificial,* Herbert Simon writes: "Only in trivial cases is the computation of the optimum alternative a trivial matter" (Simon 1996, 118). Simon points out that where optimal solutions are not known, it may well be important to look for good or satisfactory solutions rather than the best. One might add that design strategies that try to make adaptive behavior possible have to begin by assuming that the optimal behavior may not be known, or that what may seem optimal may not be.

In the fields of robotics and artificial life, one major stream of research aimed at resolving this problem of producing an adaptive and autonomous machine has focused on emergent properties and collective decisions.

Emergence is a promising yet elusive concept for the design of autonomous and adaptive robots. Instead of behavior being determined by an order from a central processor or a radio command from earth, behavior comes about as a sort of side effect of the interaction between the robot and the environment. How we should understand the nature of emergent properties and how they contribute to the production of autonomous and adaptive behavior is a matter of considerable debate. It appears that there are different kinds of emergent properties (Clarke 2001). A collective decision is one in which the interaction of the parts produces some result, rather than the result being caused by the decision of a single central processor. In this article we explore the relation among emergent properties, collective effects, and adaptability in the design of a robotic limb. In particular, we consider the goal-directed movement of a robotic arm. It is very likely that conclusions about the arm can be generalized. How far such generalizations reach is another question.

What, then, does an organism need to do to make it adaptive and autonomous in this sense of individual subtle adaptation, which may yet be very far from the fully human levels of autonomy and adaptation? While there is not a clear answer to this question, there are some illuminating guidelines from the biological world that are being applied to the design of robots. These guidelines can be enumerated as: (1) Direct Perception; (2) affordances; (3) animacy; (4) design considerations; (5) control principles; and (6) collective computational architecture.

Direct Perception

As might be expected, the manner in which a biological entity or a robot interacts with the environment is very important to its ability to adapt to perturbations of the environment. This interaction takes two forms: (1) *perception* of the environment, and (2) *action* in the environment. Perceptual psychology, specifically ecological psychology (Gibson 1966), offers a set of principles that can be used to guide robotic design toward developing a robot that interacts with its environment in ways that are indicative of how a biological system interacts with its environment. The ecological perspective views organism and environment as a single system, idealizing their relationship as one without distinct boundaries. In active perception of the world, the environment retains a direct connection to the organism, and it is the reason why the Gibsonian view of perception is called 'Direct Perception'.

Affordances

When one designs an artificial life form, environmental complexity is traditionally approached in one of two ways: (1) controlling the environmental complexity either by leaving it out of the simulation, or by creating a physical world analog that has very little complexity (often the floor of the lab), and (2) designing the organism such that it avoids, copes with, or reduces to a pile of rubble any environmental complexity that it encounters.

For example, an earthworm moving across a terrain does not do so along a straight path from point A to point B. Rather, the path of the worm is affected on a continual basis by the characteristics of the terrain, as well as by other properties of the environment through which the worm is moving (light, moisture, location of items of use to the worm, and so on) (Darwin 1881). In contrast, traditional robots and mobile agents are typically given a mobility task to accomplish, which directs them on the shortest path, independent of the characteristics of the environment, from point A to point B.

Figure 1 illustrates the hypothetical behavior of a task-based mobile robot moving through high- and low-complexity environments. In this example, the path of the task-based robot is a relatively straight line from point A to point B, independent of environmental complexity. This is because our hypothetical task-based robot is directed by its programming to go straight to the destination point. The programming constrains the robot to take the most efficient path between two points, independent of environmental irregularities, assuming that the chosen path is within the mechanical capabilities of the robot.

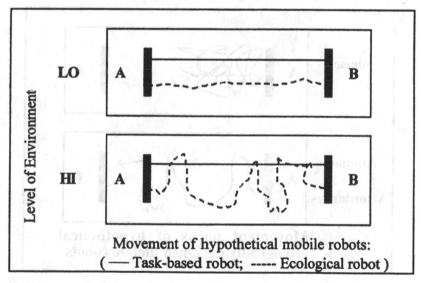

Figure 1. Comparison of point A to point B movement behavior of an ecological robot vs. a task-based robot.

Animacy

On the other hand, the movement of our hypothetical ecological robot tells a somewhat different story. This ecologically designed robot is also moving from point A to point B in the sparse environment, and doing so in a relatively straight line. Its behavior in a complexity-rich environment is completely different, however, as the characteristics of its movement are shaped by the characteristics of the terrain. This interrelationship of the organism with the environment is key to the ecological approach.

In order to modulate behavior effectively, the organism must not only be in continuous contact with the environment but must also be continuously *moving,* that is, "animate," within its environment. Close observation of a biological organism reveals a high degree of behavior, much of which seems somewhat nondirected. Indeed, biological organisms are almost constantly active and exhibit much more behavior than is "necessary" to accomplish the intent of the organism. This notion of "necessary" and "unnecessary" behavior is significant in that much of the "unnecessary" behavior has to do with being in contact with the environment, staying current, and being in position to adapt to environmental variance and take advantage of affordances. Figure 2 illustrates this hypothetical relationship.

The constant exploration of the environment used by biological systems is the necessary animation that provides the behavioral substrate for affordances and energizes the search for productive behavior solutions (Reed

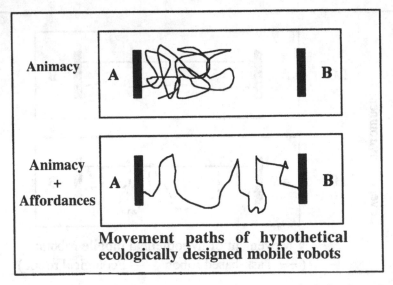

Figure 2. Comparison of point A to point B movement behavior of an ecological robot using "Animacy" vs. "Animacy + Affordances."

1996). Without animacy and the collaborating concept, affordances, there is no behavior. This has a major impact on how robots are designed! The implication is that a robot needs to be animate above and beyond the level necessary to complete its desired behavior. Reed (1996) noted that a primary characteristic of organisms is that they are animate and, as such, they search their immediate surroundings for patterns or flows of meaningful information. For example, an organism with a light sensor, such as an eye, will direct that eye at light and may move it across varying intensities of light actively to pick up information that is a particular pattern of light.

There is a distinct trade-off embedded within this concept. Animacy costs the organism energy. In the biological domain, the animacy is often associated with feeding behavior and should result in a net gain in energy. In a robot, systems resources may be overtaxed by continuous animacy. Unless, of course, the robot is only animate when there is energy available to support the animacy (cf. Mark Tilden, quoted in Menzel and D'Aluisio 2000).

Design Considerations

Machines, robots included, are typically designed to perform a specific task or tasks. In the case of an exploratory robot, the actual task is often specified as "locomotion," but in reality the task is much more. Locomotion, obstacle avoidance, use of terrain characteristics, and the like, are all aspects of the robot's behavior that are seldom specified. Indeed, some of them are nondeterministic and cannot be specified as robotic tasks. It is difficult, if not impossible, to specify a robotic task for execution in an unpredictable environment.

In nature, biological organisms seem to be not task specified but rather *behavior specified*. The difference is that task-specified robots are programmed with both "what" to do and "how" to do it, whereas biological organisms are provided with behaviors to achieve the "what" but not the details of the "how" – these are resolved in the moment that the organism *interacts* with its environment.

A side effect of this distinction is that the behavioral specifications of biological organisms are able to satisfy *many* task specifications. For example, the overhead arm motion in humans can be used for throwing a football, serving a tennis ball, or gently placing an object on an overhead shelf. Thus, specifying the arm behavior rather than the task results in a more generalized and context-specific capability for the arm.

Behavioral specification is an important issue in the design of robots that must operate in and explore the unknown. It is particularly important in its impact on control software and hardware. A more general behavior statement of robotic behavior promotes a more general approach to the control of the robot, which in turn produces more general solutions to movement situations.

A robotic device that has been behaviorally specified and is "willing" to behave in ways it was not explicitly programmed to behave must still be mechanically "competent" to move or behave in these more general ways. This sort of mechanical capability requires massively redundant mechanical degrees of freedom. Armed with more degrees of freedom than necessary to make a particular movement promotes variability in the movement. This variability in turn enhances the robot's adaptive capabilities.

Redundant degrees of freedom also provide the mechanical infrastructure necessary to allow adaptation to happen through a characteristic of human motor control known as "peripheral indeterminance." Nicholas Bernstein (1967), a Russian biomechanist, observed this phenomenon, in which humans never performed a requested arm movement exactly the same more than once. Peripheral indeterminance (also known as "motor variability") is an important capability of the human motor-control system, and much of our diversity of movement would be gone without it.

As degrees of freedom increase in a mechanical system, however, it is important to maintain some remnants of simplicity. A robot constructed from a large number of identical mechanical elements, each having one or two degrees of freedom, effectively encapsulates the complexity. In addition, the identical nature of the elements allows "polymorphism," or consistency of communication of force, rotation, velocity, and other properties among mechanical elements.

As a robotic device increases in complexity and size, it is useful to implement it *in terms of itself*. While this is a bit unusual for a mechanical system, it is quite common and beneficial in software systems. Smalltalk and LISP are two programming languages that are fundamentally implemented in themselves. In the case of a robot, this translates to designing a basic mechanical element that conforms to the design principles discussed above and then using this basic building block to assemble the actual robot. Later, we shall discuss a basic robotic-element design and illustrate how this element can be assembled into a variety of robotic systems. *The robot is not made up of parts, but rather the parts themselves are robots.*

In summary, these mechanical characteristics are critical to an effective autonomous, adaptive robotic system, but they are not the complete road map to autonomous robotic exploration. The control principles discussed in the next section will take this architectural definition a step farther.

Control Principles

It is interesting that although one realizes that biological systems exhibit complex adaptive behavior, one seldom thinks of that behavior as "control." Rather, one views control as something attributed to machines, not organisms. We are convinced that the control principles used by biological organisms can be applied to robots to produce adaptive behavior. It is important to note that the existence proof of complex, autonomous, adaptive control lies in the biological domain, *not* in any mechanical devices engineered to date. This observation leads us to search out biology's interesting and effective algorithms, rules, and representations in an attempt to use them to control autonomous, adaptive robotic systems.

An important component of the kind of robotic architecture we envision lies in its use of control rules acquired from biological examples. Biological-control researcher Kevin Kelly (1994), in a comprehensive review of a large number of complex biological and ecological systems, noted the distinctly different nature of biological control from the existing forms of control exhibited in mechanical devices. He identified four common yet critical characteristics of biological control: (1) emergent behavior; (2) redundant degrees of freedom; (3) no central director of action; and (4) a "Think Locally – Act Locally" scope of control.

The first characteristic, *emergent behavior*, is a term used in a number of disciplines to describe behavior of a system that can not be specified prior to its occurrence. This distinction between "emergent" and "specified" is critical to what sets the architecture we are proposing apart from other robotic architectures. One way of characterizing emergent properties is as "the effects, patterns or capacities made available by a certain class of complex interactions between systemic components" (Clark 2001, 114). The system here will include both the organism and the environment. The idea is that whereas the goal of a certain behavior may be specified, the way in which the goal is attained is going to emerge. We can relate emergence here to the concept of peripheral indeterminance or motor variability noted earlier. While a particular arm movement may be specified – for example, pointing – Bernstein observed that humans never performed a requested arm movement in exactly the same way more than once.

This motor variability is also related to the next characteristic exhibited by complex biological systems: *redundant degrees of freedom* in movement – that is, a greater amount of movement or behavioral variability than is necessary to complete a particular task. We previously discussed this in the context of mechanical systems, but it is important to note in addition that biological organisms generally have redundant degrees of freedom, which allow them to adapt to situations in their environments that they have never before encountered. Imagine a robot encountering a completely novel environmental situation and responding with a behavior that it was not programmed to have or that its programmers never even considered. The architecture we propose is designed to imitate the adaptability of life, and therefore includes redundant degrees of freedom with its multimodule design to provide the movement solution space necessary to produce adaptive behavior.

The third observed characteristic that some complex biological systems exhibit is *lack of a central director of action*. In most cases, the system is composed of a large number of relatively identical parts that act independently to produce complex behavior. Bird flocking and the complex behavior of ants, bees, and other social insects provide the best examples. This form of control is often preferred because of its level of redundancy and flexibility of behavioral outcome. Some of the differences between computational architectures will be detailed in the next section.

The fourth characteristic that enables biological organisms to maintain their flexibility in the face of environmental variability is *local scoping of control*. It appears that a "Think Locally – Act Locally" scoping is used by the components of many collective, complex systems. The simulation of flocking blackbirds by Craig Reynolds (1987) is a particularly clear description of potentially beneficial effects of this type of scoping in the control of a complex system. Encapsulation, an extension of this notion of local scoping of information and behavior, embraces the notion that all organisms are self-contained objects that have what they need to "get the job done." Encapsulation is apparent in biological systems, if one looks closely, and is also used in some object-oriented programming languages, Smalltalk in particular. Indeed, it appears that local scope of control is critical to the collective nature of biological control. In the architecture we propose, this style of scoping both information and behavior allows each robotic element to interact as part of a collective with a minimal amount of computational resources and pre-programmed instructions.

In sum, the architecture we propose for the design of robots intended to operate in complex environments uses all four of these principles of biological control systems. As such, it is significantly different from the architecture of the normal population of robots.

Collective Computational Architecture

Computational architectures guide the design of robotic control systems perhaps more than any other variable. Most current systems are centrally organized. Centralized control is highly insensitive to environmental change, because it relies on environmental stability. The core of its control is its own internal model of the world. Such a model is often not readily updated, because of real-time computational constraints, and is ill equipped to predict changes in the environment. Robots under the control of a centralized control architecture have only precise and highly repeatable behavior in stable environments, such as manufacturing, assembly, welding, and so on. They are not designed for adaptability, and in consequence they do not fare well when operating in the complexity we find in the real world, such as a space mission.

From a computational perspective, robotic elements in the alternative kind of architecture that we propose can be thought of as a loosely coupled collection of action systems (Reed 1982). An action system can be thought of as an autonomous animate entity that regulates its behavior according to the flow of information actively picked up from the environment (Reed 1982, 1996). At the level of an individual organism (be it biological or robotic), action systems regulate behavior with respect to meaningful cues in the environment.

These relationships can also exist between individual organisms and, in combination, can result in emergent behavior at greater levels of complexity. The organisms can be collectives; examples include bird flocks, ant colonies, and beehives, where each member constitutes an action element of the emergent behavior and each, in itself, is theoretically dispensable (Kube and Zhang 1993).

In addition, members of a collective typically do not share information with one another; rather, they interact through their local environment. Indeed, the deliberate isolation of each member's information is an important property of a collective computational architecture in that it promotes the emergence of a form of behavioral searching, resulting in a high degree of adaptation.

In sum, collective control systems operate on partial specifications. Incomplete specifications allow for, and promote, the emergence of novel movement solutions to a set of environmental conditions that could not have been pre-specified. This control characteristic is crucial to the adaptive capabilities of the architecture of an autonomous robot.

A Robotic Limb

Korienek and his colleagues have been attempting to demonstrate that these principles of robotic design derived from biological organisms actually produce adaptive behavior in robotic devices. In particular, they have been constructing a robotic arm based on the design criteria outlined above. The first stage of this project involved the construction of a puppet arm. The mechanical design of the puppet arm consisted of a collection of independent, physically constrained mechanical segments. In this arm, communication between the arm segments was limited to knowledge of relative angles of neighboring segments. The rules of a given segment used the direction and amplitude of the movement, in degrees of rotation, of the adjacent segment to determine its own movement direction and amplitude. This information was implicitly communicated through the environment, and no other information was shared between segments. The specific issue we examined was how the movement complexity in a robotic arm with 4° of freedom could be managed using a biological control rule.

Because robots were originally conceived as mechanical beings that simulated human behavior, recent conceptualizations, though mostly fictional, strive to emulate human behavior in some fashion. At the practical level the intent is to achieve movement that is as accurate, precise, and adaptable as the kind witnessed in highly skilled humans. With this goal in mind, it seems logical to base this approach on biological principles of movement while developing and implementing these control strategies.

The movement of the arm resembles the movement style of highly segmented biological appendages (such as worms, octopus tentacles, primate prehensile tails, or elephant trunks). Becuase of its collective-control architecture and highly redundant physical architecture, the arm can find multiple movement solutions to a given movement problem, thus increasing its overall adaptability and flexibility. This in turn makes it possible for the arm to be animate in the sense we specified earlier. It has many possible ways of achieving the same goal.

For this initial series of experiments, we chose what was considered the basic unit of movement at the turn of the century, the reflex arc, to model our segments. Furthermore, we applied the fundamental principle – the control strategy Facilitative Chaining – that was thought to organize these basic elements.

The puppet arm, for reasons of expediency, had no computers or motors in it. The segments were moved by a human being using the appropriate biological control rules for each segment. As the resulting data clearly displayed, although not optimally, organized goal-directed behavior can emerge from these simple principles in mechanically based as well as biologically based structures. On the other hand, the performance of this mechanical system will never be mistaken for that of the biological system it modeled. The conclusion of this study strongly suggested that the design successfully got pointing behavior to emerge from the collective decisions of the arm's segments.

As behavior emerges, it is important to assess whether specific rules allow different ways of achieving the behavioral goals they are designed to produce. Thus, in the case of the puppet arm, the sensor is a pointing sensor and guides the arm toward successful pointing. Reaching is a side effect of pointing behavior in this case. There are no rules specifically for reaching, because we did not have a reaching sensor. Some of the rules turned out to be particularly good for pointing well, even with obstacles in the way. This kind of rule represents a class of rules that are more effective at using obstacles – that is – affordances to allow pointing to occur. Such rules are very important in that they have the potential to produce generalizable behavioral solutions – in other words, adaptive behavior. Thus, in the course of this initial stage, we began to be able to distinguish rules that produce this kind of adaptive behavior from those that do not.

It is worth noting that the emergent behavior exhibited by the puppet arm possesses many of the characteristics we discussed earlier. It involves direct perception, there are no internal representations, and no memory is required. (Whether the addition of these elements would increase the adaptability of the arm is an interesting question.) The arm engages in much "unnecessary" behavior, it virtually never moves in a straight line from A to B, and, at least in the tests we conducted, it never repeated the same set of motions. In this sense, it did not achieve its goal in what one might take to be the optimal way. But in circumstances where the optimal way cannot be specified, this may turn out to be for the best. The puppet arm also succeeded in going around obstacles put in its way, in order to point at a specified target. Thus, its behavior was satisfactory in Herbert Simon's sense of that term. This behavior also suggests that the arm displayed adaptive behavior.

The puppet-arm study raised a number of interesting questions. One of these is how much complexity is required for emergent properties to emerge. (The arm had four segments.) Bird flocking, for example, seems to involve many birds. Reflection, however, suggests that Craig Reynolds's rules for bird flocking could apply just as well to a few birds as to many. There may well be kinds of emergence where this is not true. If so, the fact that in this case two or three segments may be enough to give us emergent behavior could help distinguish a particular type of emergent property from other types. Clark, for example, distinguishes between *strong* and *weak* emergence, in part, on the number of elements involved and how diverse they are (Clark 2001, 115). The pointing behavior of our robot arm might well count as weak emergence on his account. The same kind of point seems to apply to collective decisions.

The next stage in this research, as yet uncompleted, is the construction of a robotic limb with computers and motors on board. The proposed robotic limb is a mechanical device composed of between four and eight identical segments, each being elbow-shaped and oriented orthogonally at 90° to each other. Each segment can rotate in a full 360° range in 1° increments either clockwise (CW) or counterclockwise (CCW) under power of an internal motor. Figure 3 depicts a four-segment physical model of this arm.

Each segment is a self-contained unit and contains various sensors, a motor, a power supply, control electronics, and a Central Processing Unit (CPU). The rules under investigation are written in the programming language of the CPU and loaded into the on-board CPU. The operation of the segment is autonomous so long as the on-board CPU is running. Each limb segment is constructed of a variety of specially machined metal and plastic components.

Some possible basic sensory components of a given segment will include orientation in three-dimensional space (yaw, pitch, and roll), angular orientation relative to adjacent segments, and elapsed time. The segment will "sense" its local environment through these sensors. This sensed information will be incorporated into the control rules operating in the CPU of the segment. The segment will not have any other information available to its control computations. Its scope of control is such that it can sense only its own interaction with the environment and control only its own actions. Each segment acts according to information available only at the local level for example, the value of a sensor recording the change in position of a neighboring segment in the limb. Each segment acts as a single-purpose machine that encapsulates a specific adjustment to environmental change. It is this direct relationship between the animate

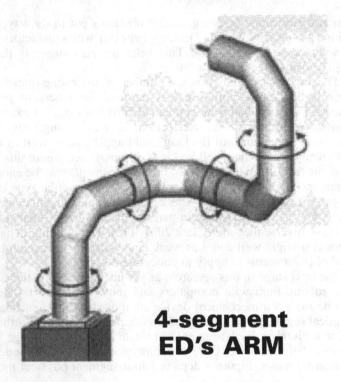

**4-segment
ED's ARM**

Figure 3.

segment and the environment that promotes the emergence of adaptive behaviors in dynamic conditions.

A multisegmented robotic limb, as in our proposed design, can be thought of as modeling a loosely coupled collection of action systems (Reed 1982). Indeed, the deliberate informational isolation of the segments from one another is an important property of the limb. This degree of separation promotes the emergence of a form of behavioral searching that results in a high degree of adaptation. In addition, it is our belief that this minimal amount of informational sharing will result in emergent behavior similar to the animal and insect collectives we mentioned earlier.

Future work in this area will expand this initial effort in an attempt to produce robotic arms that can facilitate grasping. The beneficial applications of this work are self-evident for both humans (prostheses) and machines (robots). Context-sensitive pointing, reaching, and grasping movements of the complexity and accuracy that we expect from human and robotic arms may be possible without explicit programming of the movement.

Conclusion

It is our view that a fundamental change in approach to both the control architecture and mechanical design of future robotic systems is critical to achieve the goals of full autonomy and adaptability to environmental changes. An approach involving methods and models from both the biological and the engineering sciences is needed to build this generalized robotic architecture. The analysis of adaptive behavior in terms of animacy and affordances is one important feature taken from the biological sciences that should have a major impact on how robots are designed. The implication is that the robot needs to be animate above and beyond the level necessary to complete its desired behavior. *Indeed, it is this animation that energizes the robot's search for productive behavioral solutions.* We believe that the design of the robotic arm embodies and illustrates both animacy and affordances. Its collective-control architecture and its mechanical design both contribute to these characteristics. Direct perception is an important component of autonomy, especially when the environment being explored is unknown and there is not enough time for a distant controller to respond to new information. All of these features of the design thus work together to produce a machine that behaves much more like a living thing than conventionally designed robots.

Acknowledgments

Our thanks go to the late James Gibson for the insights he has passed on to biological and computational scientists, and to the late Edward Reed for his many thoughtful extensions to the behavioral implications of ecological psychology for all life, biological and artificial.

References

Bernstein, N. 1967. *The Coordination and Regulation of Movements.* London: Pergamon Press.

Boden, M. 1996. *The Philosophy of Artificial Life.* Oxford: Oxford University Press.

Clarke, A. 2001. *Mindware.* Oxford: Oxford University Press.

Darwin, C. R. 1881. *The Formation of Vegetable Mould Through the Action of Earthworms.* London: John Murray.

Gibson, J. J. 1966. *The Senses Considered as Perceptual Systems.* Boston: Houghton Mifflin.

Kelly, K. 1994. *Out of Control: The Rise of the Neo-Biological Civilization.* Reading, Mass.: Addison-Wesley.

Korienek, G. G., A. B. Bautista, T. H. Harty, and C. Leddon. 2000. "The Use of Biologically Inspired Rules to Control a Collective Robotic

Arm." In *Distributed Autonomous Robotic Systems 4*, edited by L. E. Parker. Berlin: Springer-Verlag.

Kube, R. C., and H. Zhang. 1994. "Collective Robotics: From Social Insects to Robots." *Journal of Adaptive Behavior* 2, no. 2: 189–207.

Menzel, P., and F. D'Aluisio. 2000. *Robo Sapiens: Evolution of a New Species*. Cambridge, Mass.: MIT Press.

Reed, E. S. 1982. "An Outline of a Theory of Action Systems." *Journal of Motor Behavior* 14: 98–134.

———. 1996. *Encountering the World: Toward an Ecological Psychology*. Oxford: Oxford University Press.

Reynolds, C. W. 1993. "An Evolved, Vision-Based Behavioral Model of Coordinated Group Motion." In *From Animals to Animats 2: Proceedings of the Second International Conference on Simulation of Adaptive Behavior (SAB92)*, edited by Jean-Arcady Meyer, Herbert L. Roitblat, and Stewart W. Wilson, 384–92. Cambridge, Mass.: MIT Press.

Simon, H. 1996. *The Science of the Artificial*. Cambridge, Mass.: MIT Press.

7

A RADICAL NOTION OF EMBEDDEDNESS: A LOGICALLY NECESSARY PRECONDITION FOR AGENCY AND SELF-AWARENESS

SUSAN STUART

A partial basis for the line of argument set out in this essay is previous work by Dobbyn and myself (Stuart and Dobbyn 2000 and 2002), in which we state Kant's minimum conditions for the possibility of human conscious experience and propose that they are worthy of reconsideration – and now reinterpretation – within the contemporary debate about the mind. The Kantian paradigm, first set out in the *Prolegomena* and then expanded in the *Critique of Pure Reason*, states that we can know a priori that experience of the world will consist of a structure of spatio-temporally ordered items that are interconnected by causal laws. With this in mind, we have argued that active participation in the world, or *agency*, is possible only in a system that instantiates such a perceptual and interpretative framework, with this framework dictating how we perceive, order, and unify our experience. In this way – and in a great deal more detail – we have provided a contemporary prescriptive interpretation of Kant's descriptive metaphysical framework.

In this essay I take this idea further, examining the domain or context in which such a framework might be instantiated. Terms like *self, agent, agency, self-awareness, self-consciousness, representation, situatedness, embodied*, and, crucially for our purposes, *embeddedness*, are in frequent use in such a context, but that use is often ambiguous, inconsistent, or even just simply obscure. I shall attempt to clarify how these terms are currently used in philosophy and cognitive science, examining the limits of their use with the aim of specifying how they should be used in discourse of this kind.

Preliminary Definitions

I shall attempt to set out the conditions that an agent, whether natural or artificial, would need to satisfy for a "minimal claim to self-identity," so it will be useful to be clear from the outset what we mean by this phrase.

I use the term *self-identity* to emphasize an already existing distinction we see between the sense we have of our body and the sense we have of our self. I claim that the sense we have of our body is something that is dynamic, operating in real time, and without any conscious awareness of the operation of that sense being necessary. You might say, perhaps a little more colorfully, that our sense of body is something that we compute on the fly all the time. It is a sense that is heavily dependent on our proprioceptive ability, an ability that makes the evolution of an inner "egocentric" space possible (Brewer 1992), and is something for which we need only a bare awareness and not a conscious or self-conscious awareness – the latter being the kind of experience to which we might, in its expression or description, attach a personal pronoun. It would seem true, therefore, to say that a proprioceptive sense is essential, we might even say fundamental, to any living or artificial system that can move around in its world, sensing its world, bringing about changes in its world, and distinguishing itself from its world.[1] And by its being fundamental I mean that a proprioceptive ability or potential would have to be present in the system's neuroanatomy, or be built into the neuroanatomical equivalent in an artificial system. My claim, then, is that proprioception is a necessary, though not a sufficient, condition for a system's satisfying a minimal claim to self-identity – that is, for its being self-aware.[2]

In support of the claim to a distinction between self-consciousness and consciousness on the one hand and self-awareness on the other, it is worth pointing out, if only briefly, that the relationship between them is not one of interdependence.[3] It is certainly possible to be self-aware without being self-conscious – Leibniz recognized this in response to Descartes's claim that all experience was conscious experience[4] – but it is not, I think, clear that it is possible to be self-conscious without, at the same time, being self-aware.

Because in all cases, whether we are speaking of self-consciousness, consciousness, or self-awareness, I shall be referring to some kind of experience, I should say a few words about what I mean here by experience.

It seems true that there are at least two senses in which experience might be construed: a broad sense and a narrow sense. If experience is interpreted broadly – for example, in some Aristotelian sense – as self-awareness, it

[1] I develop this idea, alongside the six criteria for embeddedness, in another paper, "The Embedded Agent," submitted to *Minds and Machines*.

[2] In the long run it might prove more useful to describe this notion as 'agent-awareness' rather than 'self-awareness', but at present we have a proliferation of terms that need clarification, and to add a new term to the current lexicon would seem to act against this desire for clarity and simplicity.

[3] I recognize a distinction between consciousness and self-consciousness, and it is not my intention to conflate the terms, but, because I want to emphasize self-awareness as a separate experiential category and to concentrate on it, I am creating for our current purposes a binary distinction that I would not wish to defend in another context.

[4] "It is well to make distinction between the perception, which is the inner state of the monad representing external things, and apperception, which is consciousness or the reflective knowledge of this inner state . . . it is for lack of this distinction that the Cartesians have made the mistake of disregarding perceptions which are not themselves perceived" (Leibniz 1973, "Principles of Nature and of Grace," para. 4).

could be taken to encompass human, animal, and some suitably constructed machine experience. If, however, experience is interpreted narrowly as self-conscious experience, and we think of self-consciousness in the Kantian sense as requiring the ability to reflect – that it must be possible for 'I think' to accompany all my representations (Kant 1929, 152 [B131–32]) – and to offer reasons or justification for our actions, then experience encompasses, so far as we can tell, only human experience.

In this essay I interpret experience in the broad sense as self-awareness, and I endeavor to produce a set of arguments to show that the logically necessary conditions for an agent's being self-aware – satisfying a minimal claim to self-identity – can be met by specifying the necessary criteria for its being embedded.

Agents and Selves

Philosophically the term *agent* is linked to the notion of self and most frequently has been associated with ethics and philosophy of mind. An agent *acts* autonomously rather than simply *moves*, and whereas the movement of all living organisms can be explained in terms of simple, deterministic motion, the actions of human agents are generally thought not to be determined by a preset program or set of antecedent conditions. The actions of an agent are purposive within a dynamic environment, and it is argued that from this "interplay between purposive action and changing environment" (Meijsing 2000, 47) a sense of self evolves. It is this self, the Lockean *forensic* self that "belongs only to intelligent agents" (Locke 1690, bk. 2, chap. 28, §26, 220), that we think of as a moral agent. The agent behaves intentionally within its world, expressing a thought relation between itself as subject and some object. But this definition is too sophisticated to encompass the breadth of use that the term now possesses in cognitive science.

In cognitive science and artificial intelligence (AI), the terms *agent*, *situated*, and *embodied* are now widely used, in varying definitions and interpretations. In this context, an *agent* may, for example, refer to any independent piece of software for some specific task, containing special algorithms or specifically encoded knowledge that will enable it to perform that task. Examples of agents would thus be found in classical distributed systems, used to solve such problems as road-traffic monitoring (Lesser and Corkhill 1983) and air-traffic control (Cammarata et al. 1983); in Artificial Life (A-Life) systems, in which agents may evolve specialized behaviors (Aleksander 1996) or display collective behavior in concert with other agents (Mataric 1992); and in independent robotic agents that can negotiate cluttered spaces and fulfil simple goals (Brooks 1991a; Meyer and Wilson 1991). Agents seldom appear in isolation but derive most of their interesting properties from being *situated*: they exist in a world, a world that directly influences their behavior and that they, in turn, may influence. In general, agents are *embodied*: they have facilities to sense the world in which they are situated, to move through it, and, in more limited ways, to manipulate it. But these definitions are too all-inclusive to have any useful bearing on the question of what constitutes a self. In the ensuing sections, I discuss the minimal conditions for an agent's possessing any degree of self-awareness.

Agents and Their Worlds

If an artificial system is to be justifiably termed an agent, it cannot operate in isolation; it must be part of some larger system. It must perform some distinctive role in the larger system but still in some way be separable from it. By itself this is too liberal a definition: any module within a conventional data-processing system could be said, under this account, to be an agent. But such a module only receives from its environment information of a very narrowly designated kind, across narrow, serial channels, and it produces only highly stereotypical output. Furthermore, the module can have no application, or continued existence, outside the software environment for which it was designed: even modules intended for reuse generally have to be tailored for their new environments (Meyer 1996). So, as a minimum, an agent must be distinguishable from its environment yet be coupled to it in such a way that it can react to changes in it.

In Distributed AI, Artificial Life, and Robotics, agents derive much of their power and capability from being *situated*. It is hard to find any consensus on the meaning of this term, but Brooks (Brooks 1991b, 575) defines it thus: the robots "are situated in the world – they do not deal with abstract descriptions but with the here and now of the world directly influencing the behavior of the system." The agents Brooks refers to here are robots that inhabit a three-dimensional, solid, "real" world. But here we must ask an important question: whether some software agent – a virtual organism, such as a computer virus, say – receiving stimuli from a virtual world would count as being situated. In claiming that an airline-reservation system is an example of a situated system, accepting as it does messages coming into it unpredictably, Brooks seems to suggest that it would. But such a system could only very doubtfully be claimed to be situated in any sort of world: it senses nothing other than instances of the restricted class of electronic messages it can receive. So Brooks seems to mean only that the system receives and processes these messages in *real time*. Much of the thrust of Brooks's criticism of earlier robotic endeavors, such as Shakey (Nilsson 1984), is that they were able to function only in extremely artificial environments, where lighting was carefully controlled, all objects were regular polyhedra, and surfaces were painted to give maximum contrast. As we have seen, situatedness for Brooks means the functioning of an agent within the three-dimensional world of objects; but, under his own definition, an airline-reservation system cannot qualify.

A satisfactory definition of agency in the sense that the term is commonly used in A-Life and Distributed AI is probably action by a distinct, persistent, and adaptive entity performing specialized tasks in real time within a virtual software world. Although a definition of this sort might be necessary for agency, the conditions it sets out are still insufficient for an agent to which we could ascribe any form of self-awareness. Agents with even the most reduced form of self-awareness would have to be able to develop local coordinate systems, or *points of view*, locating, identifying, and interacting with objects relative to their current spatial and temporal positions. Some theorists (Suchmann 1987; Agre and Chapman 1987) refer to *situated actions*, where decisions on which action to take are based on the agent's position in the world and the state of the world at the moment the decision is to be taken. This is echoed in Brooks's insistence on the abandonment of absolute coordinate systems and the construction of distributed relational maps in which agents identify and locate objects in relation to their position at any one time. The maintenance of such a point of view is hard – though not impossible – to conceive of for any software agent in a virtual world, that is, a world without spatial dimensions.[5]

Thus, for an agent to be even minimally self-aware it must be physically distinct and extended, and be situated in a three-dimensional, real world: that is, it must be *embodied*.

A powerful strand of work in cognitive science deals with physically embodied agents, usually referred to as *animats*. Such entities are mobile and possess, at minimum, sensory systems through which they can acquire information about their environment; they may also have actuators by means of which they can manipulate their environment. Aleksander (1996) emphasises the complexity and closeness of this coupling between action, sensation, and environment. But what counts as a body in these physically embodied agents? Brooks cites an industrial paint-spraying robot as being embodied (though not situated, because it is not reactive). And now we have to ask if Aleksander's MAGNUS system could be described as embodied. Although it does possess sensors and actuators, they are of a virtual kind only and respond to a virtual world: MAGNUS exists only inside the memory of the workstation that is running it. And, problematically, under Brooks's definitions the flight-control system of an airliner would count as both situated and embodied: it receives multiple messages from sensors around the plane; it has to deal with these messages in real time; it has actuators that affect the plane's altitude, speed, and so on; and there will almost certainly be close coupling between sensors and actuators. So, an autopilot system would be a situated, embodied agent of a certain sort. But no one would want to argue that it was autonomous, or

---

[5] Although space can to some extent be simulated in software, it is doubtful whether simulation of the messy, irregular, volatile, and surprising world we inhabit is possible with current technology.

that it had sufficient conditions for purposive intelligent action, still less self-identity. Why would no one want to argue that such situated and embodied animats are not candidates for even the most minimal self-identity? Because:

- they are, in the main, reactive only: that is to say, their behavior will only be in response to detected features of, or events in, their environment;
- the goals of such systems are unitary, or else very restricted and stereotyped: an animat can be programmed to follow a wall or to retrieve and dispose of objects on the floor of a crowded office, but a richer range of possible interactions is generally impossible;
- the animats receive information from their situation across a simplified interface: although in recent work in robotics an animat may receive information from more than one type of sensor, the range of possible input data to the system is generally quite sparse;
- they have insufficient *representational* capacity.

Let us follow the last of these points for a moment and examine the issue of representation.

Representation

Many roboticists claim that the systems they build are nonrepresentational, that they contain no inner abstract description of the world and no explicit encoding of their behavior in algorithmic form. They argue that when we observe an animat negotiating some territory, or disposing of empty Coke cans, it is we, as outside observers, who *ascribe* high-level plans and intentions to it; but such plans do not exist inside the robot, which is simply reacting to its environment in a characteristic way.[6]

Claims that an animat embodies a nonrepresentational system should be examined carefully, however. There seems to be little consensus in the cognitive-science community on any definition of representation, but many approaches associate the internal states of an animat and its behavior (Wheeler 1995). For an animat A, given any internal state of A, $s(A)$, and any state in its environment, $s(E)$, explanations of representation center on the causal connection between the internal state: $s(A)$ and A's behavior, both in the presence and in the absence of $s(E)$. It is true that all such definitions are open to counterexample, but this will serve our purposes as a working definition.

Brooks's animats have been strictly engineered: the finite-state machines that govern their low-level behavior have been carefully contrived, and the patterns of connection and message passing between

---

[6] Researchers who conceptualize animats in terms of *dynamical-systems theory* make a similar claim; see, for example, van Gelder 1995.

these machines are the result of much experiment. Some machines evolve inner models of their world in the form of distributed maps (Mataric and Brooks 1990). These animats have internal states. Indeed, Brooks (1991b) does not want to claim that his robots are nonrepresentational systems; he makes only the weaker assertion that there is no *central* representation, no syntactic tokens that bear semantic weight, no variables, and no rules; internal maps are temporary and relative to the animat's current position.

The representations that the robots embody may be distributed and emergent, but they still embody representations in the sense of inner states – s(A) – that *stand for* some state of the world – s(E) – or reaction to it (A's behavior) and have causal properties. A similar point might be made against *dynamic-systems theorists*. The dynamic approach to AI conceptualizes a situated system in terms of a vector of state variables, coupled to the environment, whose values change continuously, concurrently, and interdependently over time. The internal behavior of such a system may thus be characterized as a trajectory through some high-dimensional state space (van Gelder 1998). But state vectors and state trajectories are not barred from being representational by not being symbolic; indeed, there is a considerable body of work based on investigations of the representational capacities of dynamic and chaotic systems (Freeman 1990; McKenna et al. 1994). Opponents of representational theories simply conceal a premise which states that the only form of representation is symbolic.

It has also been argued that nonrepresentationalists sidestep crucial questions about cognition: by concentrating on behavior and the control issues that arise from it, at the expense of representation and concepts, whole areas of mental life are ruled out. If an agent is going to be a candidate for self-identity, it will have to be able to synthesize and order its representations of the world from its own point of view.

Embeddedness

I shall now draw the threads of the preceding discussion together and propose a definition of embeddedness that will satisfy at least the minimal requirements for the possession of a sense of self in either a natural or an artificial agent. Embeddedness is sometimes used simply as a synonym for situatedness, at other times it is taken to refer to the close coupling between a system's intended behavior and its environment (Aleksander 1996). I believe along with other theorists (Haugeland 1993) that embeddedness can be given a more radical meaning. I would claim that the following six conditions are together necessary and sufficient for embeddedness:

1. the animat must be situated and embodied;
2. the animat must have multiple goals;

3. the environment in which the animat is situated must be sufficiently complex and challenging for the animat to be capable of complex responses to it: a robot moving across a flat plane with no obstacles would not require internal complexity of any sort to perform well; a complex and challenging environment can and should, I believe, be characterized in terms of a dynamic system, containing countless variables of significance to an animat, which are themselves coupled and which evolve over time;
4. the animat must sense its world through a rich interface: clearly the interface of an airline-booking system is impoverished in the sense that it extracts a relatively small and tightly restricted set of features from its environment;
5. the embedded animat must contain inner representations of its world for any but the most primitive sort of congnition to be possible; such representation need not be symbolic: indeed, the internal processing of the animat would almost certainly be best described in terms of a dynamic system: representations would be state vectors and trajectories through state space, and it is possible that a dynamically embedded agent would utilize internal chaotic properties for representation and memory; minimally, the agent animat will have an internal representation of a point of view on its world that is relative and shifting in response to its movement within a changing world;
6. the animat must have a rich repertoire of possible interactions with its environment, continuously manipulating its world in ways that bring about significant changes.

Given this dynamic interaction with its environment – the agent animat truly interacts, rather than simply reacts – it is possible to see the dynamic system that comprises the animat and its dynamic world as essentially parts of the same temporally evolving system.

There are, of course, problems with a radical conception of embeddedness: it is doubtful what use such a framework would be to the engineer of animats. Vectors and vector trajectories in high-dimensional spaces are notoriously difficult to work with, and the embedded conception of mind and identity is open to arguments about where the mind is actually situated – it is very hard, in such a picture, to draw a firm boundary between the mind and the world. (See, for example, Clark 1997; Clark and Chalmers 1998.) Although such objections are beyond the scope of this essay, I believe that they can be met through the concept of inner representation and the nature of the interface between an animat's internal processing and its world.

Let us now consider the notion of an agent as situated, embodied, and embedded in relation to philosophical definitions of agency and selfhood.

Embeddedness and the Self

Strawson offers a recent conception of self: "Whatever a self is, it is certainly (a) a *subject of experience*, although it is certainly (b) *not* a person, where a person is understood to be something like a human being (or other animal) considered as a living physical whole" (Strawson 1999, 99).[7] Strawson adds in an earlier essay that "any genuine sense of the self must involve a conception of the self as . . . a single, mental thing that is distinct from all other things and a subject of experience – but need not involve a conception of it as . . . an agent, or as having . . . character or personality or . . . longer-term diachronic continuity" (Strawson 1997, 424).

It seems that while cognitive science is content to have a notion of agency without self, Strawson is content to propose a notion of self without agency. In cognitive science this may be because notions of self and selfhood are inextricably bound up with ever-troublesome concepts like consciousness; but it is not clear why Strawson makes such a claim, especially when he also declares that the self is conceived of as a mental thing: the self "isn't thought of as merely a state or property of something else, or as an event, or process, or series of events. . . . It's not thought of as being a thing in the way that a stone or a cat is . . . it is thought of as something that has the causal character of a thing; something that can undergo things and do things" (Strawson 1997, 412). This would seem to be as clear a case of being an agent or having agency as one could want – acting, reacting, and interacting within one's world, influenced by the causal character of a mental thing or self; and Strawson (1999, 108) does maintain that agents are "controllers [of their environment] and intentional producers of their thoughts." Strawson cannot have it both ways. If he has self, he is committed to agency; self without agency is impossible.

I have indicated that agents must be *situated*, that is, receive and process information in real time – they must 'undergo things'; they must be *embodied*, interacting with the world, sensing and affecting change in it – they must 'undergo things and do things'. The control aspect Strawson mentions can only refer to the interactive, and therefore manipulative, participation of the agent with its world as that world appears to it from its spatially and temporally encoded point of view. I now add that they must also be *embedded*, as set out above. A conscious agent has to be able to synthesize its internal representations, the world as it appears to it, and, as Blackburn says, "be able to represent the order of [these] different appearances . . . [from] its own point of view. . . . [Thus a] minimal self-consciousness is a requirement on any kind of interpretation of experience" (Blackburn 1999, 138–39).

Both Strawson's position and the current cognitive-science position are

---

[7] In proposing this model Strawson distances himself from Bermúdez (1998, 459), who uses 'person' and 'self' interchangeably. This is not a debate that will concern us here.

untenable: if the self is conceived of as a mental thing, a thing with a 'causal character' that engages with its world in an interactive and manipulative way, it is an *agent*. I have attempted to delineate the legitimate linguistic extension of the term *agent*, and I now propose a philosophical-cum-cognitive-science notion of agent, maintaining that any minimally self-aware autonomous agent would have to be situated, embodied, and embedded, satisfying the six criteria for embeddedness, which are individually necessary but in combination are sufficient conditions for agency that is characterized by a sense of self.

References

Agre, P. E., and D. Chapman. 1987. "Pengi: An Implementation of a Theory of Activity." In *Proceedings of the Sixth National Conference on Artificial Intelligence*, 268–72. San Francisco: Morgan Kaufmann.

Aleksander, I. 1996. *Impossible Minds: My Neurons, My Consciousness*. London: Imperial College Press.

Bermúdez, J. L., A. Marcel, and N. Eilan, eds. 1995. *The Body and the Self*. Cambridge, Mass., and London: MIT Press.

Blackburn, S. 1999. *Think*. Oxford: Oxford University Press.

Brewer, B. 1992. "Self-location and Agency." *Mind* 101: 17–34.

Brooks, R. A. 1991a. "New Approaches to Robotics." *Science* 523: 1227–32.

———. 1991b. "Intelligence Without Reason." *Proceedings of the 1991 IJCAI Conference*, 569–95. '

Cammarata, S., D. McArthur, and R. Steeb. 1983. "Strategies of Cooperation in Distributed Problem Solving." In *Readings in Distributed Artificial Intelligence*, edited by A. H. Bond and L. Gasser, 102–05. San Francisco: Morgan Kaufmann.

Clark, A. 1997. *Being There*. Cambridge, Mass.: MIT Press.

Clark, A., and D. J. Chalmers. 1998. "The Extended Mind." *Analysis* 58, no. 1: 7–19.

Freeman, W. J., and Y. Yao. 1990. "Chaos in the Biodynamics of Pattern Recognition by Neural Networks." In *International Joint Conference on Neural Networks*, edited by M. Caudill, 243–46. Mahway, N.J.: Lawrence Erlbaum.

Haugeland, J. 1995. "Mind Embodied and Embedded." *Proceedings of the International Symposium on Mind and Cognition*. Taipei: Academia Sinica.

Kant, I. 1891. *Prolegomena, and Metaphysical Foundations of Natural Science*. Translated by Ernest Belfort Bax. London: George Bell and Sons.

———. 1929. *Critique of Pure Reason*. Translated by Norman Kemp Smith. London: Macmillan.

Leibniz, G. W. 1973. "Principles of Nature and of Grace" (1714). In

Philosophical Writings, edited by G. H. R. Parkinson and translated by M. Morris. London: J. M. Dent and Sons.

Lesser, V. R., and D. D. Corkhill. 1983. "The Distributed Vehicle Monitoring Testbed: A Tool for Investigating Distributed Problem-Solving Networks." *AI Magazine* 4, no. 3: 15–33.

Locke, J. 1690. *An Essay Concerning Human Understanding*. Edited by P. H. Nidditch (1975). Oxford: Oxford University Press.

Mataric, M. J. 1992. "Designing Emergent Behaviors: From Local Interactions to Intelligent Behaviour." In *From Animals to Animats 2: Proceedings of the 2nd International Conference on the Simulation of Adaptive Behaviour,* edited by J.-A. Meyer, H. Roitblat, and S. Wilson, 432–41. Cambridge, Mass.: MIT Press.

Mataric, M. J., and R. A. Brooks. 1990. "Learning a Distributed Map Representation Based on Navigation Behaviours." *Proceedings of the 1990 USA–Japan Symposium on Flexible Automation,* 499–506, Kyoto, Japan.

McKenna, T. M., T. A. McMullen, and M. F. Schlesinger. 1994. "The Brain as a Dynamical Physical System." *Neuroscience* 60, no. 3: 587–605.

Meijsing, M. 2000. "Self-Consciousness and the Body." *Journal of Consciousness Studies* 7, no. 6: 34–52.

Meyer, B. 1997. *Object-Oriented Software Construction*. Upper Saddle River, N.J.: Prentice-Hall.

Nilsson, N. J., ed. "Shakey the Robot." Stanford Research Institute AI Center Technical Report 323.

Strawson, G. 1997. "The Self." *Journal of Consciousness Studies* 4, nos. 5/6: 405–28.

———. 1999 "The Self and the SESMET." *Journal of Consciousness Studies* 6, no. 4: 99–135.

Stuart, S. A. J., and C. Dobbyn. 2000. "Kantian Descriptive Metaphysics and A-Life." In *Art, Technology, Consciousness*, edited by Roy Ascott, 194–99. Bristol, U.K. and Portland, Ore.: Intellect Books.

———. 2002. "A Kantian Prescription for Artifical Conscious Experience." *Leonardo* 35, no. 3.

Suchmann, L. 1987. *Plans and Situated Actions: The Problem of Human-Machine Communication*. Cambridge: Cambridge University Press.

van Gelder, T. J. 1998. "The Dynamical Hypothesis in Cognitive Science." *Behavioral and Brain Sciences* 21: 1–14.

Wheeler, M. 1996. "From Robots to Rothko: The Bringing Forth of Worlds." In *The Philosophy of Artificial Life*, edited by M. A. Boden. Oxford: Oxford University Press.

8

BUILDING SIMPLE MECHANICAL MINDS: USING LEGO® ROBOTS FOR RESEARCH AND TEACHING IN PHILOSOPHY

JOHN P. SULLINS

The field of robotics has always been of philosophical interest. In the past two decades the subdisciplines of behavior-based robotics, nature-based robotics, and cognitive robotics have all been particularly active; and so they offer a fascinating field of study for the empirically minded philosopher.

Largely inspired by the early work of Rodney Brooks at MIT, researchers in behavior-based and nature-based robotics seek to study small autonomous robots that do not require powerful computational capabilities to exhibit interesting behavior. Instead these robots rely on relatively simple and inexpensive machinery, with which researchers can discover how much intelligent behavior can be simulated, or instantiated, by exposing these simple computational machines to the complexity of the real world. This strategy is very different from the standard approach used by large-scale robotics research groups characterized by expensive projects funded by large universities, the Defense Advanced Research Projects Agency, or big industry. These more ambitious and well-funded projects tend to approach such standard problems in robotics such as motility, perception, and manipulation of the environment through cutting-edge programming and massive computing power that enable a robot to build complex representational models of the world from which it can formulate plans and execute actions. It is not my purpose here to criticize this approach but rather to observe that these projects are largely beyond the financial reach of the philosopher interested in robotics.

The primary characteristic of small-scale robots that makes them more appealing, from the philosopher's point of view, is that every year the necessary computational equipment needed to construct these machines gets less expensive and less technically demanding. In fact, the cost of basic equipment needed to begin exploring robotics is now easily within the meager resources of the average researcher in philosophy.

In this essay I discuss how I built a cognitive-robotics lab using inex-

Research for this paper was partially funded by a grant from the *Journal of Experimental and Theoretical Artificial Intelligence* and the philosophy department at SUNY Binghamton.

pensive LEGO® MINDSTORMS™ robot kits, and I also describe how one might build a similar lab. This lab has provided pedagogical opportunities for a number of philosophy courses, and I briefly present the results here. At the end, I put forward some ideas about the kinds of experiments that these kits make possible and how to expand beyond the limitations of the basic LEGO® hardware.

Figure 1. An example of a robot built by students to solve the "Egg Hunt Competition."

Motives

In the June 1999 issue of *Communications of the Association for Computing Machinery*, Randall Beer and his colleagues described an engineering course they teach at Case Western University using LEGO® building blocks, motors, sensors, and simple controller boards (Beer et al. 1999, 85). That article motivated me to create a robotics lab to benefit my own courses. I was scheduled to teach a course in methods of reasoning the next semester, and doing a robotics project would allow me to teach simple programming and basic mechanical reasoning in a way that would be fun yet challenging. Few teachers would argue with the observation that it is difficult to interest students in courses like critical thinking, methods of reasoning, and introduction to logic. This uphill battle can be tedious for the instructor as well, so I am always looking for ways to make the course more appealing to students.

The fundamental problem is that formal reasoning is an important part of the students' training, especially in philosophy, but most students who take these courses do so to avoid math requirements, which they perceive to be more difficult. So they are generally not in the course with the desire to learn; instead, they have the desire to avoid a "greater evil." How does

one connect with students who are already prejudiced against the formal
nature of the material taught in the course?

My answer is to make the course extremely interesting and engaging
without losing its formal component. Making formal reasoning relevant to
the students' lives and also entertaining enough to keep their attention is
the only way to achieve this goal. So I redesigned the course that I usually
teach and added a large section on simple robotics in which the students
would plan and build small robots to solve certain problems, such as navi-
gating a maze or finding and sorting different colored Easter eggs. I chose
to use the LEGO® MINDSTORMS™ robot kits that are commercially
available, as they are inexpensive and come bundled with a simple
programming language appropriate for the students with whom I was
working. Another major benefit of this approach was that it is very easy to
use other, more powerful programming languages with the LEGO® hardware, so when students were not using the machines I could use them for
experiments of my own. The LEGO® robot kits are fun and nonthreaten-
ing, but they are also capable of posing fairly complex formal problems
that require skill and understanding to solve. In fact, I have found them so
versatile that I use them in a wide range of courses beyond the methods of
reasoning class, such as foundations of AI and minds and machines.

Some examples of the philosophical issues we explored with help from
these kits in the classroom and in the lab are:

- Strengths and limitations of computationalism
- Hands-on exploration of theories in robotics, such as behavior-based
  robotics
- The Subsumption Control architecture of Rodney Brooks as a model for
  animal behavior
- Simple AI programs and the philosophical arguments counter to them
- Mind as a complex interaction of simple machines, as argued by
  Valentino Braitenberg
- Hans Moravec's transcendent-machine theory
- Peter Danielson's theories of the ethics of robotics

Equipment

Throughout the past decade the number and sophistication of do-it-your-
self robotics hobbyists and researchers outside mainstream robotics
research centers have dramatically increased. Today there are quite a few
options open to researchers interested in small-scale robotics. One can
attempt to design and build a robot from the ground up using parts available from electronics stores, hardware emporiums, and the junkyard. This
would be an exciting class project, and it is the only real option for the serious robotics researcher, but for our purposes it is too involved. For anyone
who would like to pursue this option, I recommend *The Robot Builder's*

Bonanza, by Gordon McComb (2001). This book covers a wide variety of interesting topics for the advanced designer of small robots, including, for example, servo motors, remote control, Basic Stamp, BasicX and other microcontrollers, and infrared, ultrasonic, laser, and other kinds of sensors. There is also a chapter on basic robot body construction and even a short chapter on LEGO® MINDSTORMS™ (McComb 2001).

The second option is to buy ready-made robots. There are a number of these – from the famous robotic dog Aibo to dozens of research platforms that usually take the form of small, wheeled robots with motors, touch sensors, and circuit boards prefabricated and ready to program. Some of these look quite interesting; one might simply page through the most recent issue of *Robot Science and Technology Magazine* to explore the choices. I did not select this option, because prefabricated robots tend to be quite expensive and, in the end, a bit limiting, given that one might wish to experiment with alternative sensors, such as temperature or sound sensors, that do not come with the basic robot.

The next option is a combination of the other two options: commercially available robot kits. A number of kits in the home-hobbyist market are actually quite sophisticated. The one I think best is the LEGO® MINDSTORMS™ kit released in 1998. LEGO® MINDSTORMS™ is the product of a nearly fifteen-year collaboration between a major toy company, the LEGO® Group of Denmark, and the leading think tank of technological alternatives, the MIT Media Lab (Brand 1987), in particular, the Learning and Epistemology Group (Danielson 1999, 77–83). This collaboration has resulted in a product that is useful at many levels, from pure entertainment to education and research. Such an intriguing partnership is an exciting example of bringing technical research developments from top universities to the general public. The resulting product is derived from a great deal of thought and expertise, and it is a good value. Peter Danielson (1999), in his article "Robots for the Rest of Us or for the 'Best' of Us," discusses at length the history and ethical implications of this affiliation.

The most recent version of the kit (MINDSTORMS™ 2.0) includes a 'brain brick' called an 'RCX', plus hundreds of LEGO® Technic blocks to build the body and mechanical components of a robot, two motors, two touch sensors, a light sensor, and an infrared tower, which one plugs into one's own computer to download the RCX programs one has written.

The heart of the kit is the LEGO® RCX, which contains the microcontroller that runs the robot. It is built around a Hitachi H8–series microcontroller. This is essentially a circuit board with 16KB of preprogrammed internal ROM and 32KB of static RAM. It holds the firmware, the user programs, and any data needed for the programs (Baum 2000, 15). The brick also has an LCD display and an IR (infrared) interface port to communicate with a computer that has a LEGO® IR tower (included in the kit) hooked to its COM port.

Although initially conceived as a toy, this kit has attracted the attention of many researchers and hackers in the robotics community, who quickly discovered ways to expand its out-of-the-box capabilities. Today there is a large and growing community devoted to the development of new programming environments, better sensors, and interesting experiments for the basic kit – as well as just plain showing off of the robots they have built. A rough indication of the size of this community is the search I conducted on the web using the Google search engine: it resulted in fifty-one thousand entries. The MINDSTORMS™ kit is easy to find; one can buy it in most educational-toy stores for less than $200. The kit also comes with a simple programming environment that, while limited, is remarkable in its ease of use. I found that it offered the perfect environment to introduce students who have never programmed before to the wonderful world of computer programming.

The MINDSTORMS™ software creates a graphical interface in which all the elements of the program are represented as "blocks" that stand for various commands, such as "turn on motor 'A,'" or logical conditions that enable the program to branch into alternative courses of action depending on the truth values of various sensors. These and many other pieces of the program can be "snapped" together in the virtual onscreen environment to create the final operating program, just as one would snap together LEGO® pieces in a real environment to build the robot's mechanical body.

The program is easy to use because it is impossible to create syntax errors. As a result, the programmer can concentrate on the logic of the program without the constant annoyance of errors caused by mistyping a command or some other trivial mistake. Unfortunately, this ease of use comes at a cost of functionality. The sophistication of the programs one can write is very limited. For example, one can only use a single variable, and this aspect alone severely limits what the robot can be programmed to do. Nevertheless, this and other limitations can be easily redressed by using one of the numerous alternative firmware and programming languages developed to expand the LEGO® RCX.

Currently the most popular alternative programming languages are (1) Not Quite C (NQC), a much-simplified version of the popular C programming language; (2) pbForth, a scaled-down version of the venerable Forth programming language; and (3) LegOS, replacement firmware that enables a user to program the RCX using any C compiler (Baum 2000). With these alternative programming possibilities, the MINDSTORMS™ RCX can accommodate programmers from beginners to experts.

Results

Robotics labs are far from common in philosophy departments, so when I wanted to create one I did not have any role model to work from. The courses that I used as inspiration for my lab were all based in engineering or computer-science departments. This meant that I had to alter what those courses do quite a bit and design my course with philosophy students in mind. Traditionally, courses in robotics in computer-science departments tend to be special electives that are open to advanced students who have already had a lot of computer-programming experience. I had no such luxury; the students that would take this course would be humanities majors, for the most part, with little or no computer training beyond standard word-processing applications. In addition, I would have to train them in fundamental logic and reasoning before we could dive into the robotics portion of the course. These obstacles were daunting but not impossible to overcome.

I found that with the prospect of playing with LEGO® robots, students were willing to work a bit harder on learning the basic logic needed before we could get started with working on the robots. When we reached the part of the course where we began the robotics labs, I started by assigning very simple tasks, such as "Design a robot that can wander around the room for two minutes without getting stuck."

It is surprising how difficult a simple task like this is to accomplish, as it requires an adequate mechanical solution coupled with a sound program. I divided the students into groups, making sure to spread out those few students with computer-programming ability among all the groups. There were no more than four students per robotics kit, and each group had access to a computer to program its robot. Each group planned a strategy for solving the problem and then turned that plan into reality within the time period allotted to the lab.

I wanted the labs to mimic, in a small way, the kind of atmosphere I have observed in high-tech research and development labs. In this way I could provide students with experience that might be of use later in their careers. It is my firm belief that philosophy majors have much to add to high-tech product design. It is one thing to build a high-tech gadget and quite a different problem to discover the proper fit that the gadget has in society. It is this problem that philosophy majors are equipped to solve. So one of the supplemental goals of this course was to open up some of the students to the possibility of a career in high-tech consultancy.

Each lab presented a different problem to the students, and I tried to design the lab exercises in a sequence so that by the end of the course we could have robots that could do something philosophically interesting. For instance, in the AI community there is a keen interest in robotic soccer (Wyeth 2001, 33); so in one class we built a series of robots that could solve one small part of the task of getting a ball to the other end of a field to score a goal. At the end of the class we held a competition in which teams of two robots competed against each other in a soccerlike game. I found that minor competition like this helps raise the excitement level, and it got the students to work harder than they would have otherwise.

In another class, the final project was to participate in an egg hunt,

Figure 2. A group of students work on their robot.

competing with other robots to see which one could find the most colored eggs while avoiding black eggs. An artificial life simulation was the final project in yet another class, in which competition for resources was simulated between altruistically programmed robots against egoistically programmed robots. We did try one project that I would not recommend to others: a robotic sumo competition. I selected this project because of the popularity of robot combat sports on television, and indeed it was a very popular contest with the students. The reason I consider it a failure is that it was very rough on the machines – in fact, we managed to burn out one of the ingenious little motors that come with the kits. More important than this was the general cockfight mentality that the competition generated, which

Figure 3. An example of a robot built to play soccer.

seemed very much out of place in a philosophy class. Even so, the students did learn a lot about the logic of programming, so it was not a total loss.

Building robots and programming their behavior present students with challenges that encourage creative and critical thinking. After the members of a group have thought about how to solve a problem, they must try to implement their solution. This challenge raises interesting engineering questions, as well as small-group dynamics issues that must be worked out. As a result, students develop vital skills that will serve them well in their future careers.

I have taught seven courses in three different subjects using the robotics lab. The majority of student-opinion surveys suggest that these courses are highly successful; the surveys include some of the most glowing comments I have ever received as a teacher. One student told me that technical subjects had always frightened her and so she had never considered herself capable of programming robots or entering a technical field, but after taking the course she lost such fears and was interested in looking into a technical career. Results like this make the considerable extra work of setting up and tearing down all the equipment for each lab worth it. Nevertheless, there were some negative remarks, and a small minority of students really did not take to the course at all. For some students, the paradigm shift of having to do lab work in a philosophy course was too much. Also, competition for limited computer resources made for some difficult moments, which were resolved by having some students bring in their personal laptop computers. In spite of these problems, the overall response has been so positive that I intend to continue using and expanding the lab in future courses.

How to Start a Similar Course

These courses have a number of special requirements. The most pressing need is for lab space. When I first ran the course, I had access to a small room that was, more or less, a dedicated computer lab for the philosophy department. This worked wonderfully; and I would suggest that anyone contemplating running a similar course should first solve the problem of securing adequate space and access to computers that the students can use.

I also strongly suggest keeping the course small or having student assistants help deal with little technical issues that can crop up. I found that work groups of three or four students are ideal, with one teacher or student aide for three to five groups. Each group must have access to a computer and a robotics kit, plus any extra parts needed for the specific lab assignment. One technical problem to solve before class is to set up the host computers to communicate with the RCX via an infrared transmitter, which hooks into an open COM port on the computer. The computers must have LEGO® software loaded onto them in order for this to work; so if using computers that belong to one's university, one might have to coordinate with the computing-services department to get everything set up properly in the lab.

The cost of my lab was not very high. I was able to set up the entire lab from a small grant. Each robot kit costs about $200 retail, and if one shops around one can find better deals from retailers who sell to educators. There are a number of expansion kits that add functionality to the basic set, but I would recommend avoiding these, with the exception of the "ROBOSPORTS" set, because they are not great value for the money. Eventually one will need to buy or construct additional touch and light sensors in order to do anything philosophically interesting with the robots; but this is not a hard problem to solve. One can buy a new light sensor for about $30 or construct it from parts found in an electronics store for about $3. I chose the latter option with good results. It is also vitally important to remember to budget for the considerable cost of batteries and replacement parts. Even the best students are somewhat cavalier with LEGO® pieces, and a few pieces will be destroyed inadvertently or lost each semester. I found that $5 to $10 was about the cost per kit for the damage that was caused each semester, and I solved the battery problem by investing in rechargeable batteries.

It is also important to factor in the extra preparation time necessary to teach the class. Lab exercises have to be planned and tested before the students try them. I found it useful to base some of my labs on experiments described in the books listed among the references below or on ideas I found on web sites devoted to LEGO® MINDSTORMS™. This allowed me to start using the machines right away even though I was not personally an expert in their use. It also gave me time to learn the intricacies of programming the robots so that I could then go on to design my own experiments.

One can see that building a robotics lab requires only minimal resources and a minor amount of extra work, which can be grafted onto already existing labs or tutorial centers that many philosophy departments have for teaching logic and critical thinking.

Using Small Autonomous Robots for Philosophical Research

Traditionally, philosophers have been content to talk about studies in robotics, but they have not been directly involved in fundamental research on the subject. Recently, however, there have been notable exceptions. For example, Daniel Dennett did some consulting for Rodney Brooks's ambitious Cog project, the results of which could be quite interesting.

Paradoxically, Brooks left his work on small nature-based robotics; Cog is a large, expensive robot built to resemble a human torso and is controlled by banks of powerful computers. But Brooks's move away from small robotics does not mean that there is nothing of philosophical interest left in studying simple mechanical minds. Brooks seems to have reversed his position that roboticists should not start at the top of the evolutionary scale of mental development and attempt to model humans. In his earlier work, he insisted that instead of attempting such ambitious projects we should build up our robots from the bottom of the evolutionary scale, as nature did. He felt that we should make machines that can easily maneuver around the real world and accurately sense their surroundings before we attempt to tackle poorly understood ideas like concept formation, mental mapping, and consciousness (Brooks 1991, 139).

Because philosophers are not usually the first in line to receive large grants to accomplish a project like Cog, I think that philosophers interested in robotics should begin by exploring the territory that Brooks opened up but hurriedly abandoned. Large robotics labs are motivated by engineering goals; they need to build something that works and might be marketable. Philosophers have the luxury of a more academic mind-set. We want to build machines that illustrate instances of, or point out flaws in, various theories of mind or mechanical intelligence. Roboticists know that these are topics upon which their work touches, but they tend only to footnote them in their papers or discuss them in vague or shallow ways (see Brooks 1991, 155). There is much work to do in tidying up the philosophical debris left by the last wave of robotics research.

To illustrate how this could work, let us look at three programs in North America in which interdisciplinary studies in robotics and AI research are conducted with an eye toward explicitly addressing philosophical problems. These programs are also successfully utilizing the LEGO® hardware similar to the kind I described above.

The first such program is that of the Institute for the Interdisciplinary Study of Human and Machine Cognition (IHMC) at the University of West Florida, which used the MIT 6.270 LEGO® robotics kits for an introductory course in AI. The MIT 6.270 kits were developed for a design course in electrical engineering and computer science at MIT. They are more powerful, but also more complicated, than the LEGO® MINDSTORMS™ kits I described.[1] The instructors chose these kits because they provide experience in actually doing AI. In addition,

> less obviously, experience with robotics can inform some rather difficult philosophical arguments in AI. Issues that appear intractable when discussed in the abstract seem less so in the context of autonomous robots which explore, form limited understandings of, and usefully modify a physical environment. Through the act of creating their own autonomous robots, students learn that the philosophical issues raised are interesting, and can inform, but that the objections don't prevent the real work of AI from getting done. Presenting the philosophical objections in this context acts, in a sense, as a sort of inoculation against silly, misdirected arguments that might otherwise discourage a potential AI researcher.[2]

Patrick Grim and the Group for Logic and Formal Semantics at SUNY Stony Brook have experimented with the LEGO® MINDSTORMS™ system. At the Computers and Philosophy conference at Carnegie Mellon University in August 2000, a graduate student in the Stony Brook program, Nicholas Klib, demonstrated a robot he built using the LEGO® kit. The machine was a Turing-machine simulator that flipped LEGO® blocks from a zero to a one position along a "tape" made of LEGO® pieces according to an algorithm he designed. This is a nice example of how such kits can be used to teach complex, abstract ideas in the theory of computation in a hands-on and user-friendly way.

Peter Danielson at the University of British Columbia's Center for Applied Ethics has used the LEGO® MINDSTORMS™ kits to explore issues in applied ethics.[3] As I mentioned above, Danielson wrote an outstanding review of the MINDSTORMS™ kit. He explored ethical issues raised by the kit itself and the general notion of household universal robots (Danielson 1999, 77–83). Using LEGO® robots to explore issues in ethics may not be the first thing that comes to mind when one tries to imagine the possibilities of these kits, but Danielson's work has shown that there is much to explore in this area. There are two ways in which he and his students approach ethical issues with robots. The first is by using the robots to conduct experiments simulating abstract ethical situations from

---

[1] These can be seen at the following web site: http://www.mit.edu:8001/activities/6.270/home.html.
[2] The entire statement can be read at the following web site: http://www.coginst.uwf.edu/lego/.
[3] The University of British Columbia's Center for Applied Ethics web site can be found at: http://www.ethics.ubc.ca/pad/.

which theories about the existence or evolution of altruism can be derived. The second path of research is to examine answers to the following questions: what are the proper moral attitudes to take toward robots and their creators, and what moral obligations do roboticists have to their creations and those people who will come to interact with them? This is fascinating work, which has only just begun. It is a fine example of the kinds of problems in robotics that philosophers are well trained to tackle, but which technologists might find difficult.

Conclusion

Robotics has long presented a number of fascinating philosophical issues worth exploring. Many robotics researchers acknowledge the philosophical import of their work, but they are often unskilled or unwilling to explore them. Philosophers should become more directly involved in the field of robotics; major barriers to such participation have been the expense of hardware and a lack of enough technical expertise to use it. In recent years, though, these barriers have become less problematic.

The costs of computational resources, both hardware and software, have fallen to such a low level as to place them within reach of the modest means of most philosophers. The release of LEGO® MINDSTORMS™ robotics kits in 1998 accelerated this phenomenon and afforded interested philosophers the opportunity to enter this fascinating field of study by providing all the tools needed to start researching and teaching in an environment that is fun and easy to learn. When the inevitable limitations of these kits are reached, a researcher is in a good position to branch out into more technically demanding areas of study.

As we have seen, it is not difficult to use robotics kits to develop a successful lab for research or teaching, and these kits are now being used by philosophers for pedagogical or research purposes. Such activities should be encouraged, and we can all look forward to seeing the results in both philosophy of mind and applied ethics.

References

Baum, Dave. 1999. *Dave Baum's Definitive Guide to LEGO® MIND-STORMS™* (Technology in Action). Berkeley: Apress.

———. Michael Gasperi, Ralph Hempel, and Luis Villa. 2000. *Extreme MINDSTORMS™: An Advanced Guide to LEGO® MINDSTORMS™*. Berkeley: Apress.

Beer, Randall D., Hillel J. Chiel, and Richard F. Drushel. 1999. "Using Autonomous Robotics to Teach Science and Engineering," *Communications of the ACM* 42 (June): 85–92.

Brand, Steward. 1987. *The Media Lab: Inventing the Future at M.I.T.* New York: Penguin.

Brooks, R. A. 1991. "Intelligence Without Representation." *Artificial Intelligence Journal* 47: 139–59.

Danielson, Peter. 1999. "Robots for the Rest of Us or for the 'Best' of Us." *Ethics and Information Technology* 1: 77–83.

Knudsen, Jonathan B. 1999. *The Unofficial Guide to LEGO® MINDSTORMS™ Robots*. Sebastopol, Calif.: O'Reilly.

McComb, Gordon. 2001. *The Robot Builder's Bonanza*. Second edition. New York: McGraw-Hill.

Meadhra, Michael, and Peter J. Stouffer. 2001. *LEGO® MINDSTORMS™ for Dummies*. Foster City, Calif.: IDG Books Worldwide.

Wyeth, Gordon. 2001. "RoboCup-2000: The Fourth Robotic Soccer World Championships." *AI Magazine* (March 22): 33.

9

WHAT IS THE PHILOSOPHY OF INFORMATION?

LUCIANO FLORIDI

1. Introduction

Computational and information-theoretic research in philosophy has become increasingly fertile and pervasive. It revitalises old philosophical questions, poses new problems, contributes to reconceptualising our world views, and has already produced a wealth of interesting and important results.[1] Various labels have recently been suggested for this new field. Some follow such fashionable terminology as cyberphilosophy, digital philosophy, and computational philosophy, most express specific theoretical orientations, such as philosophy of computer science, philosophy of computing or computation, philosophy of AI, computers and philosophy, computing and philosophy, philosophy of the artificial, and android epistemology. In this essay I argue that the name *philosophy of information* (PI) is the most satisfactory, for reasons that are fully discussed in section 5.[2]

Sections 2, 3, and 4 analyse the historical and conceptual process that has led to the emergence of PI. They support the following two conclusions. First, philosophy of AI was a premature paradigm, which nevertheless paved the way for the emergence of PI. Second, PI has evolved as the most recent stage in the dialectic between conceptual innovation and scholasticism. A definition of PI is then introduced and discussed in section 5. Section 6 summarises the main conclusions issuing from the preceding discussion and indicates how PI could be interpreted as a new *philosophia prima*, although not from the perspective of a *philosophia perennis*.

The view defended in this essay is that PI is a mature discipline because (a) it represents an autonomous field (*unique topics*); (b) it provides an innovative approach to both traditional and new philosophical topics (*original methodologies*); and (c) it can stand beside other branches of philosophy, offering the systematic treatment of the conceptual foundations of the world of information and of the information society (*new theories*).

---

[1] See Bynum and Moor 1998, Colburn 2000, Floridi 1999 and 2002, and Mitcham and Huning 1986 for references.

[2] The label *Philosophy of Information* was first introduced in a series of papers I have given since 1996 in Italy, England, and the United States (see note 12 below).

2. Philosophy of Artificial Intelligence as a Premature Paradigm of PI

André Gide once wrote that one does not discover new lands without consenting to lose sight of the shore for a very long time. Looking for new lands, Aaron Sloman in 1978 heralded the advent of a new AI-based paradigm in philosophy. In a book appropriately entitled *The Computer Revolution in Philosophy*, he conjectured

1. that within a few years, if there remain any philosophers who are not familiar with some of the main developments in artificial intelligence, it will be fair to accuse them of professional incompetence, and
2. that to teach courses in philosophy of mind, epistemology, aesthetics, philosophy of science, philosophy of language, ethics, metaphysics and other main areas of philosophy, without discussing the relevant aspects of artificial intelligence will be as irresponsible as giving a degree course in physics which includes no quantum theory. (Sloman 1978, 5, numbered structure added)

The prediction turned out to be inaccurate and over-optimistic, but it was far from unjustified.[3]

Sloman was not alone. Other researchers (cf., for example, Simon 1962, McCarthy and Hayes 1969, Pagels 1988, who argues in favour of a complexity-theory paradigm, and Burkholder 1992, who speaks of a "computational turn") had correctly perceived that the practical and conceptual transformations caused by ICS (Information and Computational Sciences) and ICT (Information and Communication Technologies) were bringing about a macroscopic change, not only in science but in philosophy too. It was the so-called computer revolution or "information turn." Like Sloman, however, they seem to have been misguided about the specific nature of this evolution and have underestimated the unrelenting difficulties that the acceptance of a new PI paradigm would encounter.

Turing begun publishing his seminal papers in the 1930s. During the following fifty years, cybernetics, information theory, AI, system theory, computer science, complexity theory, and ICT succeeded in attracting some significant, if sporadic, interest from the philosophical community, especially in terms of philosophy of AI.[4] They thus prepared the ground for the emergence of an independent field of investigation and a new computational and information-theoretic approach in philosophy. Until the 1980s, however, they failed to give rise to a mature, innovative, and influential

---

[3] See also Sloman 1995 and McCarthy 1995.

[4] In 1964, introducing his influential anthology, Anderson wrote that the field of philosophy of AI had already produced more than a thousand articles (Anderson 1964, 1). No wonder that (sometimes overlapping) editorial projects have flourished. Among the available titles, the reader of this essay may wish to keep in mind Ringle 1979 and Boden 1990, which provide two further good collections of essays, and Haugeland 1981, which was expressly meant to be a sequel to Anderson 1964 and was further revised in Haugeland 1997.

program of research, let alone a revolutionary change of the magnitude and importance envisaged by researchers like Sloman in the 1970s. With hindsight, it is easy to see how AI could be perceived as an exciting new field of research and the source of a radically innovative approach to traditional problems in philosophy. "Ever since Alan Turing's influential paper 'Computing machinery and intelligence' [. . .] and the birth of the research field of Artificial Intelligence (AI) in the mid-1950s, there has been considerable interest among computer scientists in theorising about the mind. At the same time there has been a growing feeling amongst philosophers that the advent of computing has decisively modified philosophical debates, by proposing new theoretical positions to consider, or at least to rebut" (Torrance 1984, 11).

AI acted as a Trojan horse, introducing a more encompassing computational/informational paradigm into the philosophical citadel (earlier statements of this view can be found in Simon 1962 and 1996, Pylyshyn 1970, and Boden 1984; more recently, see McCarthy 1995 and Sloman 1995). Until the mid-1980s, however, PI was still premature and perceived as transdisciplinary rather than interdisciplinary; the philosophical and scientific communities were, in any case, not yet ready for its development; and the cultural and social contexts were equally unprepared. Each factor deserves a brief clarification.

Like other intellectual enterprises, PI deals with three types of domain: *topics* (facts, data, problems, phenomena, observations, and the like); *methods* (techniques, approaches, and so on); and *theories* (hypotheses, explanations, and so forth). A discipline is *premature* if it attempts to innovate in more than one of these domains simultaneously, thus detaching itself too abruptly from the normal and continuous thread of evolution of its general field (Stent 1972). A quick look at the two points made by Sloman in his prediction shows that this was exactly what happened to PI in its earlier appearance as the philosophy of AI.

The inescapable interdisciplinarity of PI further hindered the prospects for a timely recognition of its significance. Even now, many philosophers are content to consider topics discussed in PI to be worth the attention only of researchers in English, mass media, cultural studies, computer science, or sociology departments, to mention a few examples. PI needed philosophers used to conversing with cultural and scientific issues across the boundaries, and these were not to be found easily. Too often, everyone's concern is nobody's business, and until the recent development of the information society, PI was perceived to be at too much of a crossroads of technical matters, theoretical issues, applied problems, and conceptual analyses to be anyone's own area of specialisation. PI was considered to be transdisciplinary like cybernetics or semiotics, rather than interdisciplinary like biochemistry or cognitive science. We shall return to this problem later.

Even if PI had not been too premature or allegedly so transdisciplinary, the philosophical and scientific communities at large were not yet ready to appreciate its importance. There were strong programs of research, especially in philosophies of language (logico-positivist, analytic, common-sensical, postmodernist, deconstructionist, hermeneutical, pragmatist, and so on), which attracted most of the intellectual and financial resources, kept a fairly rigid agenda, and hardly enhanced the evolution of alternative paradigms. Mainstream philosophy cannot help being conservative, not only because values and standards are usually less firm and clear in philosophy than in science, and hence more difficult to challenge, but also because, as we shall see better in section 4, this is the context in which a culturally dominant position is often achieved at the expense of innovative or unconventional approaches. As a result, thinkers like Church, Shannon, Simon, Turing, von Neumann, and Wiener were essentially left on the periphery of the traditional canon. Admittedly, the computational turn affected science much more rapidly. This explains why some philosophically minded scientists were among the first to perceive the emergence of a new paradigm. Nevertheless, Sloman's "computer revolution" still had to wait until the 1980s to become a more widespread and mass phenomenon across the various sciences and social contexts, thus creating the right environment for the evolution of PI.

More than half a century after the construction of the first mainframes, the development of human society has now reached a stage in which issues concerning the creation, dynamics, management, and utilisation of information and computational resources are absolutely vital. Nonetheless, advanced societies and Western culture had to undergo a digital communications revolution before being able to appreciate in full the radical novelty of the new paradigm. The information society has been brought about by the fastest-growing technology in history. No previous generation was ever exposed to such an extraordinary acceleration of technological power over reality, with the corresponding social changes and ethical responsibilities. Total pervasiveness, flexibility, and high power have raised ICT to the status of the characteristic technology of our time, factually, rhetorically, and even iconographically. The computer presents itself as a culturally defining technology and has become a symbol of the new millennium, playing a cultural role far more influential than that of mills in the Middle Ages, mechanical clocks in the seventeenth century, and the loom or the steam engine in the age of the industrial revolution (Bolter 1984). ICS and ICT applications are nowadays the most strategic of all the factors governing science, the life of society, and their future. The most developed post-industrial societies live by information, and ICS-ICT is what keeps them constantly oxygenated. And yet, all these profound and significant transformations were barely in view two decades ago, when most philosophy departments would have considered topics in PI unsuitable areas of specialisation for a graduate student.

Too far ahead of its time, and dauntingly innovative for the majority of professional philosophers, PI wavered for some time between two alternatives. It created a number of interesting but limited research niches like philosophy of AI or computer ethics – often tearing itself away from its intellectual background. Or it was absorbed within other areas as a methodology, when PI was perceived as a computational or information-theoretic approach to otherwise traditional topics, in classic areas like epistemology, logic, ontology, philosophy of language, philosophy of science, and philosophy of mind. Both trends further contributed to the emergence of PI as an independent field of investigation.

3. The Historical Emergence of PI

> Ideas, as it is said, are 'in the air'. The true explanation is presumably that, at a certain stage in the history of any subject, ideas become visible, though only to those with keen mental eyesight, that not even those with the sharpest vision could have perceived at an earlier stage. (Dummett 1993, 3)

Visionaries have a hard life. If nobody else follows, one does not discover new lands but merely gets lost, at least in the eyes of those who stayed behind in the cave. It has required a third computer-related revolution (the Internet), a whole new generation of computer-literate students, teachers and researchers, a substantial change in the fabric of society, a radical transformation in the cultural and intellectual sensibility, and a widespread sense of crisis in philosophical circles of various orientations for the new paradigm to emerge. By the late 1980s, PI had finally begun to be acknowledged as a fundamentally innovative area of philosophical research, rather than a premature revolution. Perhaps it is useful to recall a few dates. In 1982, *Time* magazine named the computer "Man of the Year." In 1985, the American Philosophical Association created the Committee on Philosophy and Computers (PAC).[5] In the same year, Terrell Ward Bynum, then editor in chief of *Metaphilosophy*, published a special issue of the journal entitled *Computers and Ethics* (Bynum 1985) that "quickly became the widest-selling issue in the journal's history" (Bynum 2000, see also Bynum 1998). The first conference sponsored by the Computing and Philosophy (CAP) association was held at Cleveland State University in

---

[5] The "computer revolution" had affected philosophers as "professional knowledge-workers" even before attracting their attention as interpreters. The charge of the APA committee was, and still is, mainly practical. The committee "collects and disseminates information on the use of computers in the profession, including their use in instruction, research, writing, and publication, and makes recommendations for appropriate actions of the Board or programs of the Association" (PAC). Note that the computer is often described as the laboratory tool for the scientific study and empirical simulation, exploration and manipulation of information structures. But then, "philosophy and computers" is like saying "philosophy and information laboratories." PI without computers is like biology without microscopes, astronomy without telescopes. But what really matters are information structures (microscopic entities, planets), not the machines used to study them.

1986. "Its program was mostly devoted to technical issues in logic software. Over time, the annual CAP conferences expanded to cover all aspects of the convergence of computing and philosophy. In 1993, Carnegie Mellon became a host site" (CAP web site).

By the mid-1980s, the philosophical community had become fully aware and appreciative of the importance of the topics investigated by PI, and of the value of its methodologies and theories.[6] PI was no longer seen as weird, esoteric, transdisciplinary, or philosophically irrelevant. Concepts or processes like algorithm, automatic control, complexity, computation, distributed network, dynamic system, implementation, information, feedback, and symbolic representation; phenomena like HCI (human-computer interaction), CMC (computer-mediated communication), computer crimes, electronic communities, and digital art; disciplines like AI and Information Theory; issues like the nature of artificial agents, the definition of personal identity in a disembodied environment, and the nature of virtual realities; models like those provided by Turing machines, artificial neural networks, and artificial life systems . . . these are just a few examples of a growing number of topics that were more and more commonly perceived as being new, of pressing interest, and academically respectable. Informational and computational concepts, methods, techniques, and theories had become powerful metaphors acting as "hermeneutic devices" through which to interpret the world. They had established a metadisciplinary, unified language that had become common currency in all academic subjects, including philosophy.

In 1998, introducing *The Digital Phoenix* – a collection of essays this time significantly subtitled *How Computers Are Changing Philosophy* – Terrell Ward Bynum and James H. Moor acknowledged the emergence of PI as a new force in the philosophical scenario:

> From time to time, major movements occur in philosophy. These movements begin with a few simple, but very fertile, ideas – ideas that provide philosophers with a new prism through which to view philosophical issues. Gradually, philosophical methods and problems are refined and understood in terms of these new notions. As novel and interesting philosophical results are obtained, the movement grows into an intellectual wave that travels throughout the discipline. A new philosophical paradigm emerges. [. . .] Computing provides philosophy with such a set of simple, but incredibly fertile notions – new and evolving *subject matters*, *methods*, and *models* for philosophical inquiry. Computing brings new opportunities and challenges to traditional philosophical activities . . . changing the way philosophers understand foundational concepts in philosophy, such as mind, consciousness, experience, reasoning, knowledge, truth, ethics, and creativity. This trend in philosophical inquiry that incorporates computing in terms of a subject matter, a method, or a model has been gaining momentum steadily. (Bynum and Moor 1998, 1)

---

[6] See, for example, Burkholder 1992, a collection of sixteen essays by twenty-eight authors presented at the first six CAP conferences; most of the papers are from the fourth one.

At the distance set by a textbook, philosophy often strikes the student as a discipline of endless diatribes and extraordinary claims, in a state of chronic crisis. *Sub specie aeternitatis*, the diatribes unfold in the forceful dynamics of ideas, claims acquire the necessary depth, the proper level of justification, and their full significance, while the alleged crisis proves to be a fruitful and inevitable dialectic between innovation and scholasticism.[7] This dialectic of reflection, highlighted by Bynum and Moor, has played a major role in establishing PI as a mature area of philosophical investigation. We have seen its historical side. This is how it can be interpreted conceptually.

4. The Dialectic of Reflection and the Emergence of PI

In order to emerge and flourish, the mind needs to make sense of its environment by continuously investing data (affordances) with meaning. Mental life is thus the result of a successful reaction to a primary *horror vacui semantici*: meaningless (in the non-existentialist sense of "not-yet-meaningful") chaos threatens to tear the Self asunder, to drown it in an alienating otherness perceived by the Self as nothingness, and this primordial dread of annihilation urges the Self to go on filling any semantically empty space with whatever meaning the Self can muster, as successfully as the cluster of contextual constraints, affordances, and the development of culture permit. This semanticisation of being, or reaction of the Self to the non-Self (to phrase it in Fichtean terms), consists in the inheritance and further elaboration, maintenance, and refinement of factual narratives (personal identity, ordinary experience, community ethos, family values, scientific theories, common-sense-constituting beliefs, and so on) that are logically and contextually (and hence sometimes fully) constrained and constantly challenged by the data that they need to accommodate and explain. Historically, the evolution of this process is ideally directed towards an ever-changing, richer, and more robust framing of the world. Schematically, it is the result of four conceptual thrusts:

(1) A metasemanticisation of narratives. The result of any reaction to being solidifies into an external reality facing the new individual Self, who needs to appropriate narratives as well, now perceived as further data-affordances that the Self is forced to semanticise. Reflection turns to reflection and recognises itself as part of the reality it needs to explain and make sense of.
(2) A de-limitation of culture. This is the process of externalisation and sharing of the conceptual narratives designed by the Self. The world of meaningful experience moves from being a private, infra-subjective

[7] For an interesting attempt to look at the history of philosophy from a computational perspective, see Glymour 1992.

and anthropocentric construction to being an increasingly inter-subjective and de-anthropocentrified reality. A community of speakers share the precious semantic resources needed to make sense of the world by maintaining, improving, and transmitting a language – with its conceptual and cultural implications – that a child learns as quickly as a shipwrecked person desperately grabs a floating plank. Narratives then become increasingly friendly because shared with other non-challenging Selves not far from one Self, rather than reassuring because inherited from some unknown deity. As "produmers" (producers and consumers) of specific narratives no longer bounded by space or time, members of a community constitute a group only apparently transphysical, in fact functionally defined by the semantic space they wish and opt (or may be forced) to inhabit. The phenomenon of globalisation is rather a phenomenon of erasure of old limits and creation of new ones, and hence a phenomenon of de-limitation of culture.

(3) A de-physicalisation of nature. The physical world of watches and cutlery, of stones and trees, of cars and rain, of the I as ID (the socially identifiable Self, with a gender, a job, a driving licence, a marital status, and so on) undergoes a process of virtualisation and distancing, in which even the most essential tools, the most dramatic experiences or the most touching feelings, from war to love, from death to sex, can be framed within virtual mediation, and hence acquire an informational aura. Art, goods, entertainment, news, and other Selves are placed and experienced behind a glass. On the other side of the virtual frame, objects and individuals can become fully replaceable and often absolutely indistinguishable tokens of ideal types: a watch is really a swatch, a pen is a present only insofar as it is a branded object, a place is perceived as a holiday resort, a temple turns into a historical monument, someone is a police officer, and a friend may be just a written voice on the screen of a PC. Individual entities are used as disposable instantiations of universals. The here-and-now is transformed and expanded. By speedily multitasking, the individual Self can inhabit ever more *loci*, in ways that are perceived synchronically even by the Self, and thus can swiftly weave different lives, which do not necessarily merge. Past, present, and future are reshaped in discrete and variable intervals of current time. Projections and indiscernible repetitions of present events expand them into the future; future events are predicted and pre-experienced in anticipatory presents; past events are registered and re-experienced in replaying presents. The nonhuman world of inimitable things and unrepeatable events is increasingly windowed, and humanity window-shops in it.

(4) A hypostatisation (embodiment) of the conceptual environment designed and inhabited by the mind. Narratives, including values, ideas, fashions, emotions, and that intentionally privileged macronarrative that is the I can be shaped and reified into "semantic objects"

or "information entities," now coming closer to the interacting Selves, quietly acquiring an ontological status comparable to that of ordinary things likes clothes, cars, and buildings.

By our de-physicalising nature and embodying narratives, the physical and the cultural are realigned on the line of the virtual. In the light of this dialectic, the information society can be seen as the most recent, although not definitive, stage in a wider semantic process that makes the mental world increasingly part of, if not *the* environment in which more and more people tend to live. It brings history and culture, and hence time, to the fore as the result of human deeds, while pushing nature, as the unhuman, and hence physical space, into the background. In the course of its evolution, the process of semanticisation gradually leads to a temporal fixation of the constructive conceptualisation of reality into a world view, which then generates a conservative closure, scholasticism.[8]

Scholasticism, understood as an intellectual typology rather than a scholarly category, represents a conceptual system's inborn inertia, when not its rampant resistance to innovation. It is *institutionalised philosophy* at its worst, that is, a degeneration of what socio-linguists call, more broadly, the internal "discourse" (Gee 1998, esp. 52–53) of a community or group of philosophers. It manifests itself as a pedantic and often intolerant adherence to some discourse (teachings, methods, values, viewpoints, canons of authors, positions, theories, selections of problems, and so on), set by a particular group (a philosopher, a school of thought, a movement, a trend) at the expense of other alternatives, which are ignored or opposed. It fixes, as permanently and objectively as possible, a toolbox of philosophical concepts and vocabulary suitable for standardising its discourse (its special *isms*) and the research agenda of the community.

In this way, scholasticism favours the professionalisation of philosophy: scholastics are "lovers" who detest the idea of being amateurs and wish to become professional. They are suffixes: they call themselves "-ans" and place before that ending (*pro-stituere*) the names of other philosophers, whether they are Aristotelians, Cartesians, Kantians, Nietzscheans, Wittgensteinians, Heideggerians, or Fregeans. Followers, exegetes, and imitators of some mythicised founding fathers, scholastics find in their hands more substantial answers than new interesting questions and thus gradually become involved with the application of some doctrine to its own internal puzzles, readjusting, systematising, and tidying up a once-dynamic area of research. Scholasticism is metatheoretically acritical and hence reassuring: fundamental criticism and self-scrutiny are not part of the scholastic discourse, which, on the contrary, helps a community to maintain a strong sense of intellectual identity and a clear direction in the

[8] For an enlightening discussion of contemporary scholasticism, see Rorty 1982, chaps. 2, 4, and especially 12.

efficient planning and implementation of its research and teaching activities. It is a closed context: scholastics tend to interpret, criticise, and defend only views of other identifiable members of the community, thus mutually reinforcing a sense of identity and purpose, instead of directly addressing new conceptual issues that may still lack an academically respectable pedigree and hence be more challenging. This is the road to anachronism: a progressively wider gap opens up between philosophers' problems and philosophical problems. Scholastic philosophers become busy with narrow and marginal *disputationes* of detail that only they are keen to ask about, while failing to interact with other disciplines, new discoveries, or contemporary problems that are of lively interest outside the specialised discourse. In the end, once scholasticism is closed in on itself, its main purpose becomes quite naturally the perpetuation of its own discourse, transforming itself into academic strategy.

What has been said so far should not be confused with the naive question as to whether philosophy has lost, and hence should regain, contact with people (Adler 1979, Quine 1979). People may be curious about philosophy, but only a philosopher can fancy they might be interested in it. Scholasticism, if properly trivialised, can be pop and even trendy – after all, "trivial" should remind one of professional love – while innovative philosophy can bear to be esoteric. Perhaps a metaphor can help to clarify the point. Conceptual areas are like mines. Some of them are so vast and rich that they will keep philosophers happily busy for generations. Others may seem exhausted until new and powerful methods or theories allow further and deeper explorations, or lead to the discovery of problems and ideas previously overlooked. Scholastic philosophers are like wretched workers digging a nearly exhausted but not yet abandoned mine. They belong to a late generation, technically trained to work only in the narrow field in which they happen to find themselves. They work hard to gain little, and the more they invest in their meagre explorations, the more they stubbornly bury themselves in their own mine, refusing to leave their place to explore new sites. Tragically, only time will tell whether the mine is truly exhausted. Scholasticism is a censure that can be applied only *post mortem*.

Innovation is always possible, but scholasticism is historically inevitable. Any stage in the semanticisation of being is destined to be initially innovative if not disruptive, to establish itself as a specific dominant paradigm, and hence to become fixed and increasingly rigid, further reinforcing itself, until it finally acquires an intolerant stance towards alternative conceptual innovations and so becomes incapable of dealing with the ever-changing intellectual environment that it helped to create and mould. In this sense, every intellectual movement generates the conditions of its own senescence and replacement.

Conceptual transformations should not be too radical, lest they become premature. We have seen that old paradigms are challenged and finally replaced by further, innovative reflection only when it is sufficiently robust to be acknowledged as a better and more viable alternative to the previous stage in the semanticisation of being. Here is how Moritz Schlick clarified this dialectic at the beginning of a paradigm shift:

> Philosophy belongs to the centuries, not to the day. There is no uptodateness about it. For anyone who loves the subject, it is painful to hear talk of 'modern' or 'non-modern' philosophy. The so-called fashionable movements in philosophy – whether diffused in journalistic form among the general public, or taught in a scientific style at the universities – stand to the calm and powerful evolution of philosophy proper much as philosophy professors do to philosophers: the former are learned, the latter wise; the former write about philosophy and contend on the doctrinal battlefield, the latter philosophise. The fashionable philosophic movements have no worse enemy than true philosophy, and none that they fear more. When it rises in a new dawn and sheds its pitiless light, the adherents of every kind of ephemeral movement tremble and unite against it, crying out that philosophy is in danger, for they truly believe that the destruction of their own little system signifies the ruin of philosophy itself. (Schlick 1979, 2: 491)

Three types of force, therefore, need to interact to compel a conceptual system to innovate. Scholasticism is the internal, negative force. It gradually fossilises thought, reinforcing its fundamental character of immobility. By making a philosophical school increasingly rigid, less responsive to the world, and more brittle, it weakens its capacity for reaction to scientific, cultural, and historical inputs, divorces it from reality, and thus prepares the ground for a solution of the crisis. Scholasticism, however, can only indicate that philosophical research has reached a stage when it needs to address new topics and problems, adopt innovative methodologies, or develop alternative explanations. It does not specify which direction the innovation should take. Historically, this is the task of two other, positive forces for innovation, external to any philosophical system: the substantial novelties in the environment of the conceptual system, occurring also as a result of the semantic work done by the old paradigm itself; and the appearance of an innovative paradigm, capable of dealing with them more successfully, and thus of disentangling the conceptual system from its stagnation.

In the past, philosophers had to take care of the whole chain of knowledge production, from raw data to scientific theories, as it were. Throughout its history, philosophy has progressively identified classes of empirical and logico-mathematical problems and outsourced their investigations to new disciplines. It has then returned to these disciplines and their findings for controls, clarifications, constraints, methods, tools, and insights. But, *pace* Carnap (1935) and Reichenbach (1951), philosophy itself consists of conceptual investigations whose essential nature is neither empirical nor logico-mathematical. To mis-paraphrase Hume: "If we take in our hand any volume, let us ask: Does it contain any abstract reasoning concerning quantity or number? Does it contain any experimental reasoning concerning matter of fact and existence?" If the answer is yes, then search elsewhere, because that is science, not yet philosophy.

Philosophy is not a conceptual aspirin, a super-science, or the manicure of language but the art of identifying conceptual problems and designing, proposing, and evaluating explanatory models. It is, after all, the last stage of reflection, where the semanticisation of being is pursued and kept open (Russell 1912, chap. 15). Its critical and creative investigations identify, formulate, evaluate, clarify, interpret, and explain problems that are intrinsically capable of different and possibly irreconcilable solutions, problems that are genuinely open to debate and honest disagreement, even in principle. These investigations are often entwined with empirical and logico-mathematical issues and so are scientifically constrained, but, in themselves, they are neither. They constitute a space of inquiry broadly definable as normative. It is an open space: anyone can step into it, no matter what the starting point is, and disagreement is always possible. It is also a dynamic space, for when its cultural environment changes, philosophy follows suit and evolves.[9] Thus, in Bynum's and Moor's felicitous metaphor, philosophy is indeed like a phoenix: it can flourish only by constantly re-engineering itself. A philosophy that is not timely but timeless is not an impossible *philosophia perennis*, which claims universal validity over past and future intellectual positions, but a stagnant philosophy, unable to contribute to, keep track of, and interact with the cultural evolution that philosophical reflection itself has helped to bring about, and hence unable to grow.

Having outsourced various forms of knowledge, philosophy's pulling force of innovation has necessarily become external. It has been made so by philosophical reflection itself. This is the full sense in which Hegel's metaphor of the Owl of Minerva is to be interpreted. In the past, the external force has been represented by such factors as Christian theology, the discovery of other civilisations, the scientific revolution, the foundational crisis in mathematics and the rise of mathematical logic, evolutionary theory, the emergence of new social and economic phenomena, and the theory of relativity, to mention just a few of the most obvious examples. Nowadays the pulling force of innovation is represented by the complex

---

[9] This normative space should not be confused with Sellars's famous "space of reasons": "In characterizing an episode or a state as that of knowing, we are not giving an empirical description of that episode or state; we are placing it in the logical space of reasons of justifying and being able to justify what one says" (Sellars 1963, 169). Our normative space is a space of design, where rational and empirical affordances, constraints, requirements, and standards of evaluation all play an essential role in the construction and evaluation of knowledge. It only partly overlaps with Sellars's space of reasons in that the latter includes more (e.g., mathematical deduction counts as justification, and in Sellars's space we find intrinsically decidable problems) and less, since in the space of design we find issues connected with creativity and freedom not clearly included in Sellars's space. For a discussion of Sellars's "space of reasons," see Floridi 1996, esp. chap. 4, and McDowell 1994, esp. the new introduction.

world of information and communication phenomena, their corresponding sciences and technologies, and the new environments, social life, and existential and cultural issues that they have brought about. This is why PI can present itself as an innovative paradigm.

5. The Definition of PI

Once a new area of philosophical research is brought into being by the interaction between scholasticism and some external force, it evolves into a well-defined field, possibly interdisciplinary but still autonomous, only if (i) it is able to appropriate an explicit, clear, and precise interpretation not of a scholastic *Fach* (Rorty 1982, chap. 2) but of the classic "ti esti," thus presenting itself as a specific "philosophy of"; (ii) the appropriated interpretation becomes an attractor towards which investigations in the new field can usefully converge; (iii) the attractor proves sufficiently influential to withstand centrifugal forces that may attempt to reduce the new field to other fields of research already well established; and (iv) the new field is rich enough to be organised in clear subfields and hence allow for specialisation.

Questions like "What is the nature of being?", "What is the nature of knowledge?", "What is the nature of right and wrong?", and "What is the nature of meaning?" are field questions. They satisfy the previous conditions, and so they have guaranteed the stable existence of their corresponding disciplines. Other questions, such as "What is the nature of the mind?", "What is the nature of beauty and taste?", and "What is the nature of a logically valid inference?" have been subject to fundamental re-interpretations, which have led to profound transformations in the definition of philosophy of mind, aesthetics, and logic. Still other questions, like "What is the nature of complexity?", "What is the nature of life?", "What is the nature of signs?", and "What is the nature of control systems?" have turned out to be transdisciplinary rather than interdisciplinary. Failing to satisfy at least one of the previous four conditions, they have struggled to establish their own autonomous fields. The question is now whether PI itself satisfies (i) to (iv). A first step towards a positive answer requires a further clarification.

Philosophy appropriates the "ti esti" question essentially in two ways, *phenomenologically* and *metatheoretically*. Philosophy of language and epistemology are two examples of "phenomenologies," or philosophies of a phenomenon. Their subjects are meaning and knowledge, not linguistic theories or cognitive sciences. Philosophy of physics and philosophy of social science, on the other hand, are plain instances of "metatheories." They investigate problems arising from organised systems of knowledge, which in their turn investigate natural or human phenomena.

Some other philosophical branches, however, show only a *tension* towards the two poles, often combining phenomenological and metatheoretical interests. This is the case with philosophy of mathematics and philosophy of logic, for example. Like PI, their subjects are old, but they have acquired their salient features and become autonomous fields of investigation only very late in the history of thought. These philosophies show a tendency to work on specific classes of first-order phenomena, but they also examine these phenomena working their way through methods and theories, by starting from a metatheoretical interest in specific classes of second-order theoretical statements concerning those very same classes of phenomena. The tension pulls each specific branch of philosophy towards one or the other pole. Philosophy of logic, to rely on the previous example, is metatheoretically biased. It shows a constant tendency to concentrate primarily on conceptual problems arising from logic understood as a specific mathematical theory of formally valid inferences, whereas it pays little attention to problems concerning logic as a natural phenomenon, what one may call, for want of a better description, rationality. Vice versa, PI, like philosophy of mathematics, is phenomenologically biased. It is primarily concerned with the whole domain of first-order phenomena represented by the world of information, computation, and the information society, although it addresses its problems by starting from the vantage point represented by the methodologies and theories offered by ICS and can be seen to incline towards a metatheoretical approach, insofar as it is methodologically critical towards its own sources.

The following definition attempts to capture the clarifications introduced so far:

(D) philosophy of information (PI) = $_{\text{def.}}$ the philosophical field concerned with (a) the critical investigation of the conceptual nature and basic principles of information, including its dynamics, utilisation, and sciences, and (b) the elaboration and application of information-theoretic and computational methodologies to philosophical problems.

Some clarifications are in order. The first half of the definition concerns philosophy of information as a new field. PI appropriates an explicit, clear, and precise interpretation of the "ti esti" question, namely, "What is the nature of information?" This is the clearest hallmark of a new field. Of course, as with any other field questions, this too serves only to demarcate an area of research, not to map its specific problems in detail (Floridi 2001). PI provides critical investigations that are not to be confused with a quantitative theory of data communication (information theory). On the whole, its task is to develop not a unified theory of information but rather an integrated family of theories that analyse, evaluate, and explain the various principles and concepts of information, their dynamics and utilisation, with special attention to systemic issues arising from different contexts of application and interconnections with other key concepts in philosophy, such as being, knowledge, truth, life, and meaning.

Recent surveys have shown no consensus on a single, unified definition of information.[10] This is hardly surprising. Information is such a powerful concept that, as an explicandum, it can be associated with several explanations, depending on the cluster of requirements and desiderata that orientate a theory (Bar-Hillel and Carnap 1953, Szaniawski 1984). Claude Shannon, for example, remarked,

> The word "information" has been given different meanings by various writers in the general field of information theory. It is likely that at least a number of these will prove sufficiently useful in certain applications to deserve further study and permanent recognition. *It is hardly to be expected that a single concept of information would satisfactorily account for the numerous possible applications of this general field.* (From "The Lattice Theory of Information," in Shannon 1993, 180)

Polysemantic concepts like information can be fruitfully investigated only in relation to well-specified contexts of use.

By "dynamics of information" the definition refers to: (i) *the constitution and modelling of information environments*, including their systemic properties, forms of interaction, internal developments, and so on; (ii) *information life cycles*, that is, the series of various stages in form and functional activity through which information can pass, from its initial occurrence to its final utilisation and possible disappearance;[11] and (iii) *computation*, both in the Turing-machine sense of *algorithmic processing* and in the wider sense of *information processing*. This is a crucial specification. Although a very old concept, information has finally acquired the nature of a primary phenomenon thanks to the sciences and technologies of computation and ICT. Computation has, therefore, attracted much philosophical attention in recent years. Nevertheless, PI privileges "information" over "computation" as the pivotal topic of the new field because it analyses the latter as presupposing the former. PI treats computation as only one (although perhaps the most important) of the processes in which information can be involved. Thus, the field should be interpreted as a philosophy of information rather than just of computation, in the same sense in which epistemology is the philosophy of knowledge, not just of perception.

From an environmental perspective, PI is prescriptive about, and legislates on, what may count as information, and how information should be

---

[10] For some reviews of the variety of meanings and the corresponding different theoretical positions, see Braman 1989, Losee 1997, Machlup 1983, NATO 1974, 1975, 1983, Schrader 1984, Wellisch 1972, Wersig and Neveling 1975. I have defended a revision of the definition of semantic information as meaningful data in Floridi (forthcoming).

[11] A typical life cycle includes the following phases: occurring (discovering, designing, authoring, etc.), processing and managing (collecting, validating, modifying, organising, indexing, classifying, filtering, updating, sorting, storing, networking, distributing, accessing, retrieving, transmitting, etc.), and using (monitoring, modelling, analysing, explaining, planning, forecasting, decision making; instructing, educating, learning, etc.).

adequately created, processed, managed, and used.[12] PI's phenomenological bias, however, does not mean that it fails to provide critical feedback.
On the contrary, methodological and theoretical choices in ICS are also
profoundly influenced by the kind of PI a researcher adopts more or less
consciously. It is therefore essential to stress that PI critically evaluates,
shapes, and sharpens the conceptual, methodological, and theoretical basis
of ICS – in short, that it also provides a *philosophy of ICS*, as this has been
plain since early work in the area of philosophy of AI (Colburn 2000).

It is worth stressing here that an excessive concern with the metatheoretical aspects of PI may lead one to miss the important fact that it is perfectly
legitimate to speak of PI even in authors who lived centuries before the
information revolution. Hence it will be extremely fruitful to develop a
historical approach and trace PI's diachronic evolution, so long as the technical and conceptual frameworks of ICS are not anachronistically applied
but are used to provide the conceptual method and privileged perspective to
evaluate in full the reflections that were developed on the nature, dynamics,
and utilisation of information before the digital revolution (consider, for
example, Plato's *Phaedrus*, Descartes's *Meditations,* Nietzsche's *On the
Use and Disadvantage of History for Life*, and Popper's conception of a
third world). This is significantly comparable with the development undergone by other philosophical fields, like philosophy of language, philosophy of biology, and philosophy of mathematics.

The second half of the definition indicates that PI is not only a new field
but provides an innovative methodology as well. Research into the conceptual nature of information and its dynamics and utilisation is carried on
from the vantage point represented by the methodologies and theories
offered by ICS and ICT (see, for example, Grim, Mar, and St. Denis 1998).
This perspective affects other philosophical topics as well. Information-
theoretic and computational methods, concepts, tools, and techniques have
already been developed and applied in many philosophical areas, to extend
our understanding of the cognitive and linguistic abilities of humans and
animals and the possibility of artificial forms of intelligence (philosophy
of AI; information-theoretic semantics; information-theoretic epistemology; dynamic semantics); to analyse inferential and computational
processes (philosophy of computing; philosophy of computer science;
information-flow logic; situation logic); to explain the organizational principles of life and agency (philosophy of artificial life; cybernetics and
philosophy of automata; decision and game theory); to devise new
approaches to modelling physical and conceptual systems (formal ontology; theory of information systems; philosophy of virtual reality); to
formulate the methodology of scientific knowledge (model-based

---

[12] Following this research, Herold 2001 has defined librarianship as applied philosophy
of information and has suggested that "librarianship, as an applied PI, seems to be the last
of all disciplines to traverse intellectual history and digest its paradigms."

philosophy of science; computational methodologies in philosophy of science); and to investigate ethical problems (computer and information ethics; artificial ethics), aesthetic issues (digital multimedia/hypermedia theory; hypertext theory and literary criticism), and psychological, anthropological, and social phenomena characterising the information society and human behaviour in digital environments (cyberphilosophy). Indeed, the presence of these branches shows that PI satisfies criterion (iv). As a new field, it provides a unified and cohesive theoretical framework that allows further specialisation.

PI possesses one of the most powerful conceptual vocabularies ever devised in philosophy. This is because we can rely on informational concepts whenever a complete understanding of some series of events is unavailable or unnecessary for providing an explanation. In philosophy, this means that virtually any issue can be rephrased in informational terms. This semantic power is a great advantage of PI understood as a methodology (see the second half of the definition). It shows that we are dealing with an influential paradigm, describable in terms of an informational philosophy. But it may also be a problem, because a metaphorically pan-informational approach can lead to a dangerous equivocation – namely, thinking that since any x can be described in (more or less metaphorically) informational terms, the nature of any x is genuinely informational. And the equivocation obscures PI's specificity as a philosophical field with its own subject. PI runs the risk of becoming synonymous with philosophy. The best way of avoiding this loss of identity is to concentrate on the first half of the definition. PI as a philosophical discipline is defined by what a problem is (or can be reduced to be) *about*, not by *how* the problem is formulated. Although many philosophical issues seem to benefit greatly from an informational analysis, in PI information theory provides a literal foundation, not just a metaphorical superstructure. PI presupposes that a problem or an explanation can be legitimately and genuinely reduced to an informational problem or explanation. So the criterion to test the soundness of the informational analysis of x is not to check whether x *can* be formulated in information terms but to ask what it would be like for x not to have an informational nature at all. With this criterion in mind, I have provided a sample of some interesting questions in Floridi 2001.

6. Conclusion: PI as *Philosophia Prima*

Philosophers have begun to address the new intellectual challenges arising from the world of information and the information society. PI attempts to expand the frontier of philosophical research not by putting together pre-existing topics, and thus reordering the philosophical scenario, but by enclosing new areas of philosophical inquiry – which have been struggling to be recognised and have not yet found room in the traditional philosophical syllabus – and by providing innovative methodologies to address

traditional problems from new perspectives. Is the time ripe for the establishment of PI as a mature field? We have seen that the answer can be affir-
mative because our culture and society, the history of philosophy, and the
dynamic forces regulating the development of the philosophical system
have been moving towards it. But then, what kind of PI can be expected to
develop? An answer to this question presupposes a much clearer view of
PI's position in the history of thought, a view probably obtainable only a
posteriori. Here it might be sketched by way of guesswork.

We have seen that philosophy grows by impoverishing itself. This is
only an apparent paradox: the more complex the world and its scientific
descriptions turn out to be, the more essential the level of the philosophical discourse understood as *philosophia prima* must become, ridding itself
of unwarranted assumptions and misguided investigations that do not
properly belong to the normative activity of conceptual modelling. The
strength of the dialectic of reflection, and hence the crucial importance of
one's historical awareness of it, lies in this transcendental regress in search
of increasingly abstract and more streamlined conditions of possibility of
the available narratives, in view not only of their explanation but also of
their modification and innovation. How has the regress developed? The
scientific revolution made seventeenth-century philosophers redirect their
attention from the nature of the knowable object to the epistemic relation
between it and the knowing subject, and hence from metaphysics to epistemology. The subsequent growth of the information society and the
appearance of the infosphere, the semantic environment in which millions
of people spend their time nowadays, have led contemporary philosophy
to privilege critical reflection first on the domain represented by the
memory and languages of organised knowledge, the instruments whereby
the infosphere is managed – thus moving from epistemology to philosophy
of language and logic (Dummett 1993) – and then on the nature of its very
fabric and essence, information itself. Information has thus arisen as a
concept as fundamental and important as being, knowledge, life, intelligence, meaning, and good and evil – all pivotal concepts with which it is
interdependent – and so equally worthy of autonomous investigation. It is
also a more impoverished concept, in terms of which the others can be
expressed and interrelated, when not defined. In this sense, Evans was
right:

> Evans had the idea that there is a much cruder and more fundamental concept
> than that of knowledge on which philosophers have concentrated so much,
> namely the concept of information. Information is conveyed by perception, and
> retained by memory, though also transmitted by means of language. One needs
> to concentrate on that concept before one approaches that of knowledge in the
> proper sense. Information is acquired, for example, without one's necessarily
> having a grasp of the proposition which embodies it; the flow of information
> operates at a much more basic level than the acquisition and transmission of
> knowledge. I think that this conception deserves to be explored. It's not one

that ever occurred to me before I read Evans, but it is probably fruitful. That also distinguishes this work very sharply from traditional epistemology. (Dummett 1993, 186)

This is why PI can be introduced as the forthcoming *philosophia prima*, both in the Aristotelian sense of the primacy of its object – information – which PI claims to be a fundamental component in any environment, and in the Cartesian-Kantian sense of the primacy of its methodology and problems, as PI aspires to provide a most valuable, comprehensive approach to philosophical investigations.

PI, understood as a foundational philosophy of information design, can explain and guide the purposeful construction of our intellectual environment, and it can provide the systematic treatment of the conceptual foundations of contemporary society. It enables humanity to make sense of the world and construct it responsibly, a new stage in the semanticisation of being. Clearly, PI promises to be one of the most exciting and fruitful areas of philosophical research of our time. If what has been argued in this essay is correct, its current development may be delayed but is inevitable, and it will affect the overall way in which we address both new and old philosophical problems, bringing about a substantial innovation of the philosophical system. This will represent the information turn in philosophy.[13]

References

Adler, M. 1979. "Has Philosophy Lost Contact With People?" *Long Island Newsday*, 18 November.

Anderson, A. R. 1964. *Minds and Machines*. Contemporary Perspectives in Philosophy Series. Englewood Cliffs: Prentice-Hall.

Bar-Hillel, Y. 1964. *Language and Information*. Reading, Mass., and London: Addison-Wesley.

[13] I discussed previous versions of this article on many occasions: at the University of Bari, where I was invited to give a series of lectures entitled *Epistemology and Information Technology* during the 1996–1997 academic year, and then at the University of Rome III, when I gave a course on the philosophy of information as visiting professor in epistemology, in 1999–2000; at the Applied Logic Colloquium (Queen Mary and Westfield College, London University, 26 November 1999); at Computing in Philosophy: One-day colloquium on the philosophical implications of computing and its uses and consequences for philosophical studies (King's College, London University, 19 February 1999); at the 1999 APA Eastern Division Meeting, special session arranged by the APA Committee on Philosophy and Computers (Boston, 28 December 1999); and at the Fourth World Multiconference on Systemics, Cybernetics, and Informatics (Orlando, Fla.: 23–26 July 2000). I am very grateful to the organisers of these meetings and the participants for their feedback. Charles Ess, Jim Fetzer, Ken Herold, and Jim Moor read the final version and made very valuable comments. All remaining errors are mine. This research has been partly supported by two grants, one from the Coimbra Group, Pavia University, and one from Wolfson College, Oxford.

Bar-Hillel, Y., and Carnap, R. 1953. "An Outline of a Theory of Semantic Information." Reprinted in Bar-Hillel 1964, 221–74; page references are to this edition.

Boden, M. A. 1984. "Methodological Links between AI and Other Disciplines." In *The Study of Information: Interdisciplinary Messages*, edited by F. Machlup and V. Mansfield. New York: John Wiley and Sons. Reprinted in Burkholder 1992.

———. 1990. *The Philosophy of Artificial Intelligence*: Oxford Readings in Philosophy. Oxford: Oxford University Press.

Bolter J. D. 1984. *Turing's Man: Western Culture in the Computer Age*. Chapel Hill: University of North Carolina Press.

Braman, S. 1989. "Defining Information." *Telecommunications Policy* 13: 233–42.

Burkholder, L., ed. 1992. *Philosophy and the Computer*. Boulder, San Francisco, and Oxford: Westview Press.

Bynum, T. W. 1998. "Global Information Ethics and the Information Revolution." In Bynum and Moor 1998, 274–89.

———. 2000. "A Very Short History of Computer Ethics." *APA Newsletters on Philosophy and Computers* 99, no. 2 (Spring).

———, ed. 1985. *Computers and Ethics*. Oxford: Blackwell. Published as the October 1985 issue of *Metaphilosophy*.

Bynum, T. W., and Moor, J. H., eds. 1998. *The Digital Phoenix: How Computers Are Changing Philosophy*. Oxford: Blackwell.

CAP. Web site of the Computing and Philosophy annual conference series, http://www.lcl.cmu.edu/caae/cap/CAPpage.html.

Carnap. R. 1935. "The Rejection of Metaphysics." Chapter in *Philosophy and Logical Syntax*. Reprinted 1997. Bristol: Thoemmes Press.

Colburn, T. R. 2000. *Philosophy and Computer Science*. Armonk, N.Y., and London: M. E. Sharpe.

Dummett, M. 1993. *Origins of Analytical Philosophy*. London: Duckworth.

Floridi, L. 1996. *Scepticism and the Foundation of Epistemology: A Study in the Metalogical Fallacies*. Leiden: Brill.

———. 1999. *Philosophy and Computing: An Introduction*. London and New York: Routledge.

———. 2001. "Open Problems in the Philosophy of Information." The Herbert A. Simon Lecture on Computing and Philosophy, CAP meeting, Carnegie Mellon University, 10 August 2001, http://caae.phil.cmu.edu/caae/CAP/.

———. Forthcoming. "Is Information Meaningful Data?". Preprint available at http://www.wolfson.ox.ac.uk/~floridi/pdf/iimd.pdf.

———, ed. 2002. *The Blackwell Guide to the Philosophy of Computing and Information*. Oxford: Blackwell.

Gee, J. P. 1998. "What is Literacy?" In *Negotiating Academic Literacies:*

Teaching and Learning Across Languages and Cultures, edited by V. Zamel and R. Spack, 51–59. Mahwah, N.J.: Erlbaum.

Glymour, C. N. 1997. *Thinking Things Through: An Introduction to Philosophical Issues and Achievements.* Cambridge, Mass.: MIT Press.

Grim, P., G. Mar, and P. St. Denis. 1998. *The Philosophical Computer.* Cambridge, Mass.: MIT Press.

Haugeland, J. 1981. *Mind Design: Philosophy, Psychology, Artificial Intelligence,* Montgomery, Vt.: Bradford Books.

————. 1997. *Mind Design II: Philosophy, Psychology, Artificial Intelligence,* Cambridge, Mass.: MIT Press.

Herold, K. R. 2001. "Librarianship and the Philosophy of Information." *Library Philosophy and Practice* 3, no. 2. Available at http://www.uidaho.edu/~mbolin/lppv3n2.htm.

Kuhn, T. S. 1970. *The Structure of Scientific Revolutions.* Second edition. Chicago: University of Chicago Press.

Losee, R. M. 1997. "A Discipline Independent Definition of Information." *Journal of the American Society for Information Science* 48, no. 3: 254–69.

Machlup, F. 1983. "Semantic Quirks in Studies of Information." In *The Study of Information: Interdisciplinary Messages*, edited by F. Machlup and U. Mansfield, 641–71 New York: John Wiley.

McCarthy, J. 1995. "What Has AI in Common with Philosophy?" Proceedings of the 14th International Joint Conference on AI. Montreal, August 1995. Available at http://www-formal.stanford.edu/jmc/aiphil.html.

————, and P. J. Hayes. 1969. "Some Philosophical Problems from the Standpoint of Artificial Intelligence." *Machine Intelligence* 4: 463–502.

McDowell, J. 1994. *Mind and World.* Cambridge, Mass: Harvard University Press.

Mitcham, C., and A. Huning, eds. 1986. *Philosophy and Technology II: Information Technology and Computers in Theory and Practice.* Dordrecht and Boston: Reidel.

NATO. 1974. Advanced Study Institute in Information Science, Champion, 1972. *Information Science: Search for Identity*, edited by A. Debons. New York: Marcel Dekker.

————. 1975. Advanced Study Institute in Information Science, Aberystwyth, 1974. *Perspectives in Information Science*, edited by A. Debons and W. J. Cameron. Leiden: Noordhoff.

————. 1983. Advanced Study Institute in Information Science, Crete, 1978. *Information Science in Action: Systems Design*, edited by A. Debons and A. G. Larson. Boston: Martinus Nijhoff.

PAC. Web site of the American Philosophical Association Committee on Philosophy and Computers, http://www.apa.udel.edu/apa/governance/committees/computers/.

Quine, W. V. O. 1979. "Has Philosophy Lost Contact with People?" *Long Island Newsday*, 18 November. The article was modified by the editor. The original version appears as essay no. 23 in *Theories and Things*. Cambridge, Mass.: Harvard University Press, 1981.

Reichenbach, H. 1951. *The Rise of Scientific Philosophy*. Berkeley: University of California Press.

Ringle, M. 1979. *Philosophical Perspectives in Artificial Intelligence*. Atlantic Highlands, N.J.: Humanities Press.

Rorty, R. 1982. *Consequences of Pragmatism*. Brighton: Harvester Press.

Russell, B. 1912. *The Problems of Philosophy*. Oxford: Oxford University Press.

Schlick, M. 1979. "The Vienna School and Traditional Philosophy." English translation by P. Heath. In *Philosophical Papers*. 2 volumes. Dordrecht: Reidel. Originally published in 1937.

Schrader, A. 1984. "In Search of a Name: Information Science and Its Conceptual Antecedents." *Library and Information Science Research* 6: 227–71.

Shannon, C. E. 1993. *Collected Papers*. Edited by N. J. A. Sloane and A. D. Wyner. Los Alamos, Calif.: IEEE Computer Society Press.

Simon H. A. 1962. "The Computer as a Laboratory for Epistemology." First draft. Revised and published in Burkholder 1992, 3–23.

———. 1996. *The Sciences of the Artificial*. Cambridge, Mass.: MIT Press.

Sloman A. 1978. *The Computer Revolution in Philosophy*. Atlantic Highlands, N.J.: Humanities Press.

———. 1995. "A Philosophical Encounter: An Interactive Presentation of Some of the Key Philosophical Problems in AI and AI Problems in Philosophy." Proceedings of the 14th International Joint Conference on AI. Montreal, August 1995. Available at http://www.cs.bham.ac.uk/~axs/cog_affect/ijcai95.text.

Stent, G. 1972. "Prematurity and Uniqueness in Scientific Discovery." *Scientific American* (December): 84–93.

Szaniawski, K. 1984. "On Defining Information." Now in Szaniawski 1998.

———. 1998. *On Science, Inference, Information and Decision Making: Selected Essays in the Philosophy of Science*, edited by A. Chmielewski and J. Wolenski. Dordrecht: Kluwer.

Torrance, S. B. 1984. *The Mind and the Machine: Philosophical Aspects of Artificial Intelligence*. Chichester, West Sussex, and New York: Ellis Horwood Halsted Press.

Wellisch, H. 1972. "From Information Science to Informatics." *Journal of Librarianship* 4: 157–87.

Wersig, G., and U. Neveling. 1975. "The Phenomena of Interest to Information Science." *Information Scientist* 9: 127–40.

10

THE SUBSTANTIVE IMPACT OF COMPUTERS ON PHILOSOPHY: PROLEGOMENA TO A COMPUTATIONAL AND INFORMATION-THEORETIC METAPHYSICS

RANDALL R. DIPERT

It is now acknowledged that the vast majority of philosophers in the United States, perhaps 90 percent or more, make considerable use on a daily basis of personal computers for their work. This could be for word processing, professional e-mail, literature searches, or downloading journal articles and original texts. There are also far more substantive influences that the computer age may be having, or will have, on philosophy. After a brief survey of practical uses of computers in philosophy, I shall turn to the influence that the widespread familiarity with digital computers has had on debates in the philosophy of mind. Then I shall sketch what are, I believe, the more important and novel past and future influences of computation- and information-based ideas on metaphysics.

Pedagogical and Practical Uses of Computers

A majority of philosophers would undoubtedly swear to the importance of word processing for their writing and of e-mail for philosophical discussions, or at least for professional communication of some sort with their colleagues, such as internal departmental administrative matters, conference and publishing details, requests for articles, and so on. Many of us subscribe to Internet discussion lists connected to philosophy, although I no longer do. Open discussion lists often become chatty, heatedly argumentative, bloated, or otherwise demonstrate a variant of Gresham's Law that bad money drives out good. It has proven difficult to choose suitable editors for edited discussion lists, or even to find someone willing to take on the task – which often requires daily intervention, for no remuneration.

In pedagogy, the availability of PCs, now for almost every college student, and access to the Internet have been much praised. The benefits of Internet access might include distance learning or, more widely accepted as a step forward, access to some or all class material online: texts, assignments, syllabus, and the like. A positive contribution to pedagogy might also include individual student interchange with the instructor, which displaces traditional and notoriously sparse office hours. For many students, e-mail is less intimidating than other forms of exchange. (However, from the instructor's point of view, verbal inadequacies in e-mail may be more disappointing than in face-to-face conversations, and the familiarities and colloquialisms common in students' personal e-mail may be seriously disagreeable to instructors and poor preparation for later, more formal encounters with workplace superiors or fellow workers through writing.) I myself draw the line at chatting and various forms of instant messaging with students, as I suspect most instructors do.

Discussion groups and chatting among students in the same course about course content add another possibility to the computer's enhancements for educational communication. I do not think such discussion has proven itself to be useful, however, either by empirical studies or by the quantity of anecdotal reports favoring or disfavoring this option. No doubt, too, its value would depend on the course at hand. There are arguments to be made for such chat groups as an ad hoc discussion section for an introductory course on philosophical issues or history, while for logic they might do more harm than good by spreading incorrect conceptions or techniques. Logic is a field where one cannot get much satisfaction or enjoyment out of just expressing one's views and obtaining some feedback from them. Many of these enhancements for pedagogy – as well as word processing and e-mail – would have been technologically feasible with a great deal less computing power than we now have on our desks, and in some cases with virtually no "computing" at all. However, the availability of personal computers, spurned by a fad for them, has made such communication cheap, as a matter of economy of scale, and easy, as compared to hard-wired word processing, noncomputerized electronic data storage, individual teletype machines, and multiway telephoning.

The realm of logic pedagogy is a very special case that has been considered elsewhere, as has the ability to use tools for assisting reasoning and organized writing in a natural language. These issues are considered in articles in this collection and in *The Digital Phoenix*.

The usual discussion of computers and philosophy centers on the impact or influence of computers and the theory of computers on philosophy. A less often told story is the influence of philosophy, especially logic, on the development and theory of computers. It is unlikely that advances in computers could have taken place so rapidly without certain theoretical notions having already been in place. For example, the early developments in electrical switching (by relays) and then electronic switching (vacuum tubes and transistors) relied for their theory upon the algebraic system of a universe of discourse with only True/False (or 1/0), explored by George Boole in the late 1840s and early 1850s. The even more important analysis of "logic gates," such as AND, NOR, NAND, and so on, and accurately predicting the results of wiring together many of these gates, relied upon such notions having already become well known.[1] Even Boole was not the first to develop the logics we associate with Boolean Logic (Dipert 1993), although his became the best known such work. By contrast, the logically more exact and deeper developments by Gottlob Frege, Bertrand Russell, Alfred North Whitehead, and so on, using a predicate calculus and quantifiers, seem to have had no impact on computer hardware design.

The unification of Boolean binary semantics with a binary number system for numerical calculation was the idea of the German engineer Konrad Zuse in the 1930s and became common after World War II. Simultaneous with the more pronounced commercial efforts to develop rapid and accurate numerical calculating machines (from Babbage, even from Pascal and Leibniz, through the 1940s and 1950s) were sustained efforts to build "logic machines" that could assess the validity of arguments or even produce valid conclusions from given premises (Dipert 1999). Some of these "philosophical" efforts were known and integrated into computer development; others were not. One odd fact is that the first electrical design for any sort of computing or calculating device seems to have been sketched by the philosopher and logician C. S. Peirce in the 1890s. The very notion of "decidability" was articulated by David Hilbert in 1900 (if it didn't exist before, see Dipert 1984), in his famous list of unsolved problems, and was improved upon in later works of the 1920s and 1930s, concluding with Church's Theorem in 1938.

Substantive Influences of Computers on Philosophy

Among the more substantive contributions that our increasing awareness of the nature and potential for computation can make to philosophy, two broad strands of "computer impact" stand out. First, there is a far greater awareness of precise concepts and theorems in information theory and computation theory. I say "awareness" because the precise concepts were actually developed, and the main theorems proven, before the advent of high levels of computing power – say, enough power to determine deducibility- or validity-relationship among propositions in standard propositional logic (which arrived in 1955). The sharpening of such basic conceptions of computation as decidability, the idealized notion of a computer in the Turing Machine, and, more broadly, a theory of recursive functions are almost entirely products of the 1920s and 1930s. Information theory and the precise measurements of quantitative information we are now so familiar with (baud-rate, memory storage in terms of kilobytes and

---

[1] Boolean constructs are also helpful in doing library and literary searches, as well as for Internet search engines. The careful use of Boolean constructions, however, seems to be less, not more, supported (or hidden as an "option") as use of these searching tools spreads over a general population.

megabytes) likewise arose from the work of Claude Shannon in the 1940s. Shannon was exclusively interested in the mathematization of ideal signal-carrying capacity of "channels" (mainly wires and electromagnetic radiation) for long-distance electronic communication.

The second strand of computer influence on philosophy is more evident in philosophical debates that are still very active (although they also began with Turing or earlier). This is the extent to which computers and symbol-transforming devices make possible an extension of the Cartesian "mechanization" of the material world, to mechanical (lawlike) theories of the mind. While some of these issues were of course already posed by Descartes, the sheer visibility of personal computers, some successes in artificial intelligence, and the widespread use of mathematics-assisting programs, chatbots, and so on, have all made the problem more acute even for laypersons. Imagining or building computers to perform complex symbolic tasks, such as language processing and inference, made it begin to seem possible to view the mind, too, as a kind of machine. If the solar system (and then all matter) is like a machine, running with clocklike regularity according to clearly expressible laws, then is our mind (maybe brain) a machinelike entity running according to symbol or semantic-information manipulating laws?

However, connecting the formulation of precise, mathematical representations of the world (while using the suggestive word *mechanics*) to what is "mechanical" and then connecting this notion to the root metaphor of "machines" has many disanalogies that are often overlooked. It is true that sophisticated, intricate, small, multipart, machined-metal devices arose first (or were widely distributed) at the end of the Renaissance and the beginning of the Modern period: firing mechanisms for guns, telescopes, microscopes, clocks, and locks, for example. No doubt there is *some* connection or causal association with the rise of modern physics (chemistry was "mechanized" in the nineteenth century, parts of biology only in the past sixty years). But the disanalogies are clear. Machines are subject to wear, metal fatigue, or wood deterioration, and to the use of flawed materials, irregular crystallization, or impurities. The "mechanistic" conception of the physical world has no such flaws. It is difficult to imagine what would be analogous to machines' imperfections applied to nature: perhaps aging and changing physical constants. Another disanalogy is that machines are manmade. They are artifacts intentionally crafted, designed, or altered according to a conception of their purpose or purposes. The conception of God as clockmaker that so amused Hume or motivated William Paley is only an attempt at a kind of integrity with this robust mechanistic image – namely, by giving nature purposiveness just as machines have. Those who developed mechanical views without purposiveness proved unwilling to bite the bullet in order to retain a so-called mechanical conception of the world. Instead, they invented the idea of purely natural "laws" and perfect regularities without further justification.

"Law," like "machine," is just another imperfect metaphor. When stripped of the trappings of its beautiful and impressive mathematics, a mechanical conception hardly seems to be an "explanation" of the ways of the world as hitherto sought.[2] Nature is somehow "like" a machine and governed by basic principles somehow "like" the Mosaic laws of God or the laws of man.[3]

The application of a mechanistic worldview to human beings is nicely summarized by Julien de la Mettrie's succinct phrase and book title, "L'homme machine." But the view harks back to Hobbes, if not earlier. In Hobbes's case this clearly includes a mechanistic (and in fact physicalistic) account of mental events.

It is a puzzling fact in the history of ideas (especially in the sixteenth, seventeenth, and twentieth centuries) that the successes of the mathematical/mechanistic conception of the world and explanation of some observable phenomena have been collapsed with *physical* mechanism. It is easy to imagine the possibility of perfect, mathematically describable laws that are not (necessarily) reduced to physicalistic-mechanistic principles. There might exist laws of mental phenomena whose relationship to ultimate physical laws is left open. Likewise, one might have admitted that Kepler's and especially Newton's accounts of the behavior of the solar system were perfectly mechanistic/mathematical, while admitting that they may not be the ultimate such explanation (to corpuscles, to the Special and General Theories of Relativity, particle and quantum mechanics, and so on). So too one might admit that there are perfect laws of mental phenomena. These could be scientific-empirical, mathematical, and mechanistic but (so to speak) lie at a higher or different level of description, which may, or may not, be susceptible to further reduction. One might remain at this explanatory level because one views the further reduction as (for now) unhelpful or nonexplanatory, or in order not to beg questions (the reduction of mind to brain) that seem, even today, matters more of faithlike belief or guesswork than science or even philosophy.

Our best thinking on these topics is based on some promising research still in an infant stage (for example, the Churchlands' "neurological" though not yet fully physicalized explanations). Or it is based on complicated, and possibly question-begging, examples that no one can reasonably claim are fully convincing and logically correct and complete. These are the tentative arguments against such reductions, and they include the

[2] Even with perfect regularities described with mathematical formulae, we do not know whether this is a basic or a higher-level phenomenon, itself explained by still more fundamental perfect regularities – the Periodic Table in terms of quantum chromodynamics, for example.

[3] As a consequence of the hundreds of years of use of these metaphors (starting in the sixteenth century, accelerating in the seventeenth, and remaining very much with us in the twenty-first), these metaphors have hardened and no longer trouble us, as they probably should.

thought experiments of Nagel's bat, Searle's Chinese Room, and Jackson's and Chalmers's Mary, inverted spectra, and zombies. It should not be the solemn task of philosophers merely to guess at which reduction or explanation seems promising, or will succeed, or will not. Philosophical reflections have traditionally aimed at better-established and even necessary truths, not at mere intuitions or hopes that might be useful for guiding us in deciding which research proposal to fund.

New Horizons and "Bizarre" Metaphysics

For most contemporary philosophers who have entered the fray, the ultimate physical nature of mentality is demonstrable if and only if mental activity can be explained as computation. That is, the overlap is total between those who are physicalists and those who believe that all mental states and processes are to be explained computationally (computationalism). It is of course possible that some physicalist philosophers have not thought much about computation and the distinctive issues in the philosophy of mind. This could have been because the computational turn in the philosophy of mind had not yet taken place (Hobbes) or because it was not well explored (Quine). It took some time before Gödel's Incompleteness Theorems were widely viewed as having little or nothing to say about human minds and formal computation.[4]

One thing we can do is to break apart the two claims, mentation-as-computation and all-as-physical. We could be a physicalist and yet reject the usually implemented physical computations as sufficient for mentation. This seems to be Searle's position. All is physical for him, whether this is hope, faith, useful assumption for science, or rational belief. Yet mentation is not precisely what we usually mean by computation, or is implemented in a fundamentally different way from the artificial processors and systems of semiconductors we now have.

Another position is that mentation is computation, but that not all is physical. That is, we might suppose that another monistic "basic substance" implements, or could implement, computation (or, less interestingly, that something like Cartesian dualism is true). This path is one that David Chalmers (in Chalmers 1996) explores as Proto-panpsychism and, somewhat differently later in the volume, in the suggestion of an information-content and an information-transformation metaphysics that is yet to be worked out.[5] Such views cannot precisely be, say, Leibnizian

---

[4] See the historical discussion of the impact of Gödel's Theorems in Simon 1992, 3–5. An exception is Penrose 1989.

[5] The intriguing information-structure view has not been much discussed in the extensive secondary literature and criticism that Chalmers's work has generated. He himself has not seriously pursued these alternative metaphysics, so far as I know. A fascinating account of panpsychism that is motivated by recent philosophy of mind, especially Chalmers's views, is Gregg Rosenberg's in Rosenberg 1997. This does not connect to the information-

monads, as monads are "windowless" – they have no causal interaction. Something like causation, or a notion of changing ontological dependencies, would be necessary for there to be "transformations" of information.

Example 1: Eric Steinhart's Digitalization of the World and Its Lawlike Behavior
One of the first of the more unusual applications of thinking in the digital age to metaphysics is in Steinhart 1998. A similar view is advanced in the more readable but less detailed Steinhart 1994. Steinhart makes several basic claims in the 1998 article:

1. Metaphysics is the study of the foundations of physics (p. 1).[6]
2. Ultimate reality is a massively parallel computing machine sufficiently universal for the realization of any physically possible world (p. 1).[7]
3. "Ultimate reality is computational space-time. . . . [This system of ideas is called] *digital metaphysics*" (p. 1).
4. "[P]rograms are [simply?] *orderings of abstract transformations of abstract states of affairs*" (p. 3).
5. States of affairs are constituted by states at discrete points in space-time: naturally phenomena are only ideally *continuous* – for their utility in our conceptualization and mathematization of nature – but are really *discrete* (p. 121, paraphrased and extended).
6. "Nature is only finitely complex" (p. 122).
7. "All things [earlier: "some appearances"] are patterns of monads over some set of monads."

While the invoking of the language of metaphysics, including such terms as *monads*, is new and the research program – as philosophy – is bold, the basic computational model is not. It is roughly a conception of the world as cellular automata, an idea that dates from von Neumann and Stanislaw Ulam in the late 1940s. Cells, with discrete states, are arranged in a space. The cells and their states seem roughly to correspond to Steinhart's monads, while the laws governing their transformations are Steinhart's

theoretic metaphysics I incline to, and it leaves open the question of what experience is "about," or what the varieties of all experience are. It is nevertheless a highly developed view that is not the usual uninspired milling about.
  [6] The plural of "foundation" is curious, since surely we should entertain the possibility of, and maybe even persistently strive for, *one* foundation. The politically preferred language of social toleration and diversity seems to have struck everywhere.
  [7] It is unclear why this structure of ultimate reality has to be both "massive" and involve "parallel processing." If not infinite, then a computer with a single-processing von Neuman–style architecture can compute whatever a parallel processing computer can (this is a proven theorem). Likewise, as I demonstrate in Dipert 1997, a massive amount of complexity and hence information can, using relations, be embedded in rather small abstract structures (e.g., graphs with forty vertices). I suspect Steinhart is being unduly influenced by the superficial appearance of the physical world as we now conceive of it.

programs.[8] Such data elements and programs are well-known, as implemented in The Game of Life, Civilization, and so on. There is, however,
already a difficulty. For if the cells or monads are themselves arranged in
a relativistic space-time, then what is the "order" mentioned in the notion
of "programs," and in what sequence do the transformations induced by
the programs take place? The ordinary notion of a program, or rather of the
running of a program,[9] would put these state transformations in a separate
notion of "time." That is, if the cells are arranged in a space-time, this time
is not identical to the order of actions, transformations, that the program
performs.

In Steinhart's admirably succinct presentation, roughly in the style of
the metaphysics of Leibniz's *Monadology* and Wittgenstein's *Tractatus*
(although without numbers in Steinhart's original presentation), we are
left to wonder what states of affairs are, what data types are computed or
transformed, and what precisely does the computing – both what is or
contains the "program" and what does the "running" of it (and in what
"time" this takes place). Although perhaps inspired by such ideas as
cellular automata, Steinhart's view clearly extends and self-consciously
develops a metaphysics beyond CAs, as they are called. By integrating a
digitalized notion of space-time (which would be necessary for digital as
opposed to analog computation) and in its abstract formulation of
programs as "orderings of transformations," it is extremely imaginative.
Steinhart also connects each of these ideas to literature and movements
in contemporary physics.[10]

In other respects, Steinhart's views are likely to be misleading to
philosophers and to be conservative ("classical") within physics itself.
From the point of view of etymology and cognates, it is natural enough to
describe what physics talks about, or perhaps the foundational elements of
what physics talks about, as "the physical." What counts as the basic
elements of physical reality for physicists, however, would often make
philosophers choke, or at least confuse them. In this sense, the basic
elements of what physics studies are *not* "physical" in the philosophers'
sense. I mean to allude to the "color" of quarks and other metaphorical
properties, nonlocality or indeterminacy, probabilistic waves, and so on.
Philosophers' physicalism, insofar as any philosophers articulate a clear

[8] But compare Steinhart's note 7, p. 127: "Monads are individual computing entities."
This makes it sound as though monads are what *do* computations, rather than *upon what* the
computing is done (usually, "data").

[9] Chalmers 1996 uses a similar distinction to make an insightful critique of Searle's
Chinese Room.

[10] Another imaginative point in Steinhart's conception is that his monads are not
"substantial" particulars: they "aren't *made out* of any kind of stuff" (n. 7, p. 127). He calls
them *"functional* particulars," although this sounds like a pragmatic-positivistic turn in
metaphysics that is driven by epistemology, as are other points, e.g., the continuous nature
of space and time "are certainly not empirically warranted" (p. 121). Neither is its discrete
nature, except by equally dubious *ad ignorantium* arguments.

conception of it, more often than not contains strong elements of pre-twentieth-century materialism (and thus classical mechanics).

Example 2: Randall R. Dipert's World as Pure Structure
In an article from almost the same period and without any influence between the two, I developed a theory (Dipert 1997) that is in some ways similar to Steinhart's but in other ways divergent.[11] As examples of their affinities, I explicitly propose that the "deepest" layer of reality is discrete and implicitly argue that these deepest phenomena are large and combinatorially are horrendously complex, although ultimately finite. I too was inspired by cellular automata and contemporary literature in physics. (Steinhart was braver than I was: he has sections entitled "Against Natural Actual Infinities" and "Nature Is Only Finitely Complex," Steinhart 1998, 121–23). Both of us explore the nature of pure patterns (which I call "structures").

My own view is this. An ideal description of the structure of the world, what physics and cosmology aim for, is necessarily a mathematical structure. Mathematical structures require at least one relation in order to be nontrivial. In fact, it is possible to describe any structure at all in terms of structures using only one symmetric relation, which I dub the "World Relation." I argue that a subset of these graphs is maximally simple, in terms of the numbers of properties and relations needed to describe a structure, any structure. The structures induced by one symmetric relation are precisely those that Graph Theory describes. Undirected Graph Theory is a theory of only combinatorically possible arrangements of vertices, also called "nodes," and arcs; or, if you will, the structures built only from connecting lines and points, where length of the line is irrelevant. Remarkably, even small such structures – with seven to forty nodes – carry enormous amounts of information because of the phenomenon of combinatorially explosive nonisomorphic graphs. With only nine nodes, there are 308,708 distinct, that is, nonisomorphic, graphs (Dipert 1997, 345, from Harary and Palmer 1973). Using a Leibnizian metaphysical principle of "no distinction without a difference," I further reduce this number to only nonautomorphic graphs; yet even with this limitation, the number of distinct automorphic graphs, and thus "simplest" information-carrying entities, rises suprapolynomially. Such a feature is necessary if we guess that the universe is, information-theoretically, enormously complex and has a colossal number of other such possible structures, that is, possible worlds. Likewise, the physical world and its basic principles appear very difficult to analyze; this is mirrored by the NP completeness of various properties of graphs, such as algorithms for discerning isomorphisms in subgraphs.

In several respects, however, Steinhart is "classical" in his physics, while

---

[11] I have since had conversations and correspondence with Steinhart.

I include radical speculation. For example, I do not assume that space and time and space-time are fundamental, or that a foundational account of them is necessary for a theory to count as physics. Although I am not an idealist with respect to space and time, believing for example as Kant and perhaps Leibniz did, space and time are nevertheless structures that are, to some extent, impositions of finite minds on experience. I see no reason other than hoary tradition, however, to suppose that space and time *are* fundamental to any, or "the," correct and complete account of reality. They might be embedded in deeper structural features of reality; that is, their patterns may be reducible to, or in that dubious language "supervene on," more basic structure. It is as if we can see some higher-level regularized grid – or *almost* perfectly regularized grid – in more complex patterns. Thus, I am neither an idealist nor an "eliminitivist" with respect to space and time.

For me, everything is structure, internally or externally defined.

I perhaps went overboard in two respects. First, I idolized graphs as the epitome of pure structure and, second, and relatedly, I assert the superiority of metaphysics whose sole property is one two-place relation (and symmetric at that, to match the vertices of undirected graphs).[12] What remains correct, I believe, is that we should understand finite relational combinatorics as the primary vehicle of all information, and that this focus (on metamathematical descriptions of all structure as well) allows us to give information-theoretic arguments for graphs (or other mathematical structures) as basic, minimal data-structures. I should have been more careful to say that Graph Theory is not the only mathematical theory that accomplishes what I wish. There are many bridge theorems between Group and Graph Theory, for example, suggesting that Group Theory could also serve as a fundamental theory. Important graph-theoretical results, such as Polyá's Enumeration Theorem, appear to rely crucially on group-theoretic/algebraic reasoning.

The variety of nonisomorphic and nonautomorphic graphs are, I would conjecture, the only minimal and perfect characterization of entities that convey information. They "are" not themselves pure informational enti-

---

[12] That is, I rule out graphs or basic structural arrangements from being asymmetric or transitive. "Directed" graphs are those that have a direction to the connectedness of two nodes, usually represented notationally by an arrow rather than a line segment. Directedness seems to be a stronger hypothesis than is needed, as directedness can be represented by undirected basic structural connections (and vice versa), and the theorems concerning directed graphs are brought at a higher price and thus are far less known and studied. I am interested in the weakest – least information-carrying – single relation that suffices for the diversity and complexity of a world; information is carried out by the nature of the resulting distinct structures rather than by constraints on the basic relation itself. A further exploration of these issues would require a digression into Coding Theory (how to encode and recover information in diverse optimal ways, such as overcoming noise, reducing the required Baud rate and so on). By these means, we could more carefully connect Information Theory and Graph theory, although Coding Theory is most interested in information-theoretic properties within linear strings of discrete, finite elements, such as 1s and 0s.

ties, but they are an ideal, or an almost ideal, representation of information-carrying entities of all sorts.

Conclusion

It is clear that the effects of computers on philosophical work habits in the past decades have been massive. It is far less clear that the quality of philosophizing has been enhanced. More substantively, the theory of computation has led to detailed arguments concerning a computational model of all forms of mental activity, of mentation. This would have been possible, however, well before computers were ubiquitous, as Turing himself demonstrated. Recent philosophy of mind has essentially treated the computational theory of mind as a corollary of physicalism. To this extent, it has not been especially interesting, as it just gives added detail to an already well-explored worldview or faith.

I do not believe that "computational philosophy" has yet yielded its most interesting and imaginative metaphysical implications. The obvious taking of sides in the profession and the standard demarcation between physicalists and nonphysicalists, as well as merely stomping around well-worn places on the dancefloor, do not exhaust the philosophical implications of the computational turn. There are very probably radical new conceptions of reality that can yet be squeezed out of pure, abstract conceptions of reality as information, structure, and information-transformation. An information-theoretic metaphysics has been mentioned by David Chalmers and developed by Eric Steinhart, and a metaphysics of pure structure has been developed by Randall R. Dipert, motivated by information-theoretic arguments. To both of these ends we would need very general but precise notions of (mathematical) structure of all sorts, and a precise qualitative, as opposed to Shannon's quantitative, approach to information.

References

Chalmers, David J. 1996. *The Conscious Mind: In Search of a Fundamental Theory.* Oxford: Oxford University Press.

Dipert, Randall R. 1984. "Peirce, Frege, Church's Theorem, and the Logic of Relations." *History and Philosophy of Logic* 5: 49–66.

———. 1993. "History of Modern Logic," macropedia article "Logic." In *Encyclopedia Britannica*, all editions since 1993.

———. 1997. "The Mathematical Structure of the World: The World as Graph." *Journal of Philosophy* 94, no. 7: 329–58.

———. 1999. "Logic Machines and Diagrams." In *Routledge Encyclopedia of Philosophy*.

Harary, Frank, and E. M. Palmer. 1973. *Graphical Enumeration.* New York: Academic Press.

Penrose, Roger 1989. *The Emperor's New Mind: Concerning Computers, Minds, and the Laws of Physics*. Oxford: Oxford University Press.

Rosenberg, Gregg 1997. "A Place For Consciousness: Probing The Deep Structure of The Natural World." Book MS based on Indiana University Ph.D. dissertation of 1997. Available online at http://www.ai.uga.edu/~ghrosenb/book.html

Simon, Herbert A. 1992. "The Computer as a Laboratory for Epistemology." In *Philosophy and the Computer*, edited by Leslie Burkholder, 3–23. Boulder, Colo.: Westview Press.

Steinhart, Eric. 1994. "Structural Idealism." *Idealistic Studies* 24, no. 1: 77–105.

———. 1998. "Digital Metaphysics." In *The Digital Phoenix*, edited by Terrell Ward Bynum and James H. Moor, 117–34. Oxford and Malden, Mass.: Blackwell Publishers.

11

COMPUTATION AND CAUSATION

RICHARD SCHEINES

In 1982, when computers were just becoming widely available, I was a graduate student beginning my work with Clark Glymour on a Ph.D. thesis entitled "Causality in the Social Sciences." Dazed and confused by the vast philosophical literature on causation, I found relative solace in the clarity of Structural Equation Models (SEMs), a form of statistical model used commonly by practicing sociologists, political scientists, and so on to model causal hypotheses with which associations among measured variables might be explained. The statistical literature around SEMs was vast as well, but Clark had extracted from it a particular kind of evidential constraint first studied by Charles Spearman at the beginning of the twentieth century, the "vanishing tetrad difference."[1] As it turned out, certain kinds of causal structures entailed these constraints, and others did not. Spearman used this lever to argue for the existence of a single, general intelligence factor, the infamous g (Spearman 1904).

In 1982, we could, with laborious effort, calculate the set of tetrad constraints entailed by a given SEM. We did not, however, have any general characterization of the connection between qualitative causal structure, as represented by the "path diagram" for a SEM (Wright 1934), and vanishing tetrad constraints. If we could only find such a characterization, we thought, then we could lay down a method of causal discovery from statistical data heretofore written off as impossible. Two or three times a week I would come in to Clark's office offering a conjecture, which, as is his wont, he would immediately claim to refute with a complicated counterexample. Although Clark is a man of astounding vision and amazing intellectual facility, he could never be accused of being fastidious in calculation, and thus I would dispute his counterexample on the grounds that he had calculated incorrectly (I had no other advantage over him). We

I thank Emily Scheines and Martha Harty for patient reading and wise counsel.

[1] A vanishing tetrad difference is an equality among the products of correlations involving an entire foursome of variables. For example, a vanishing tetrad difference among W, X, Y, Z is $\rho_{wx}*\rho_{yz}=\rho_{wy}*\rho_{xz}$. Such a constraint is implied, for example, by a model in which there is a single common cause of W, X, Y, and Z. In Spearman's case, tetrad differences among measures of reading and math aptitude led him to hypothesize that a single common cause, general intelligence, was responsible for performance on all four psychometric instruments.

would then work through the example several times, each time getting a different answer. After the better part of an hour we would sometimes converge on a calculation we could both endorse, but the process was so laborious as to make progress toward the larger goal almost hopeless.

Finally, after a particularly long and mind-numbing session, Clark said to me, "Why not write a computer program that would do these calculations for us? The algorithm for computing the vanishing tetrad difference for a given model is clear enough. You're young, you can still learn a new trick or two." Having not the faintest idea of how many late-night hours I would spend debugging code over the next several years, I went out and bought a book on Pascal and plunged in. Peter Spirtes joined us a few years later, Kevin Kelly helped out for a few years before he went off to apply formal learning theory to epistemology, and together Clark, Peter, and I have computationally attacked the epistemology of causation for nearly twenty years. By 1984, with the help of the crude program I had written and another from Kevin, we had developed an automatic procedure for correcting a given SEM. By 1987 we had a graphical characterization of when a SEM entails a vanishing tetrad difference as well as a different (but related) empirical regularity, the vanishing partial correlation (Glymour, Scheines, Spirtes, and Kelly 1987). In 1988, because we had become involved in the artificial-intelligence community, we became aware of Judea Pearl's work on Bayes Networks. Combining our work on causal discovery, which came from the linear causal model tradition, with Pearl's, which came from computer science, produced a perspective that was much more fertile than the sum of its parts.

In what follows I try to survey this synthesis. To leave the story accessible, I neglect formality and detail wherever I can; where I cannot I try to minimize it, and at the end I point the way to four sources that have all the detail one could want. I begin by sketching the philosophical perspective on the subject that has dominated discussion for more than two thousand years. I then sketch the work in biological and social science on linear causal modeling over the past century, which fed directly into my own. I next describe the work in computer science that took place almost independently of linear causal modeling. After discussing the synthesis between linear causal modeling and computer science, I sketch the enormous progress in causal epistemology and algorithmic causal discovery this synthesis unleashed. I think it is not in the least an overstatement to say that the computational turn has radically and permanently changed the philosophical, computational, and statistical view of causation.

Causal Analysis before the Computer

For nearly two thousand years the philosophical analysis of causation has emphasized reducing causal claims, for example, "A is a cause of B," to more primitive or well-understood concepts. In this section I give a whirlwind account of these attempts and try to indicate convincingly that none has succeeded.

Following Hume's famous regularity theory, J. L. Mackie (1974) gave an account in which causes are INUS conditions for their effect, that is, Insufficient but Necessary parts of Unnecessary but Sufficient sets of conditions. Although logical relations are clear enough, they are not up to the task of capturing even simple features of causation without excessive ad hocery – for example, the asymmetry of causation, or the distinction between direct and indirect causation.

Hume and Mackie also gave an analysis of causation in terms of counterfactuals, but the most systematic and sophisticated counterfactual theory of causation is from David Lewis (1973). According to Lewis, event A was a cause of event B if A occurred, B occurred, and there is no possible world in which A does not occur but B does that is closer to the actual world than one in which A does not occur and B does not occur either. Building a semantics for causation on top of similarity metrics over possible worlds is a dubious enterprise, but even if one likes possible worlds it seems clear that Lewis has it backward. We make judgments about what possible worlds are more or less similar to the one we inhabit on the basis of our beliefs about causal laws, not the other way around. Causal claims support counterfactual ones, but not vice versa. Further, Lewis attempts to capture the asymmetry of causation with "miracles," but the attempt fails.

Philosophers have also tried to reduce causal relations to probabilistic ones. In Patrick Suppes's 1970 theory, A is a prima facie cause of B if A occurs before B in time and A and B are associated. Variables A and B are probabilistically dependent, if for some value b of B, $P(A) \neq P(A \mid B = b)$. We notate independence between variables A and B as $A \perp\!\!\!\perp B$, and association as $A \not\perp\!\!\!\perp B$. A is a *genuine cause* of B if A is a prima facie cause of B and there is no event C prior to A such that A and B are independent conditional on C, that is, $A \perp\!\!\!\perp B \mid C$.

First, why probability should be considered less mysterious than causation is a mystery to me. Second, this theory has us quantify over all possible events C prior to A, a requirement that makes the epistemology of the subject hopeless. Third, the probabilistic theory requires temporal knowledge. Fourth, the theory rules out cases in which A is a cause of an intermediary I, I is a cause of B, A is also a direct cause of B, but the influence of A on B through I is opposite in sign and of exactly the same strength as that of A on B direct, leaving A and B independent and thus apparently not a cause of B according to this and other theories like it.

In figure 1, for example, if the direct effect of the tax rate on tax revenues (positive) was the exact same strength as the indirect effect of tax rate on revenue through economic activity (negative, because the effect of tax rate on economic activity is negative and the effect of economic activity on tax revenue is positive, which combine to produce a negative effect), then Tax Rate $\perp\!\!\!\perp$ Tax Revenues.

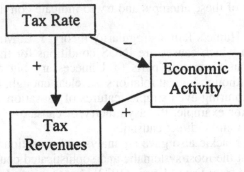

Figure 1. Canceling causes

The only remaining candidate is a manipulability theory, in which A is a cause of B just in case there are two distinct manipulations of A that result in some difference in B (its value or its probability, for example). Manipulability theories account well for the asymmetry of causation. Just because two different manipulations of A result in a difference in B does not mean that two different manipulations of B would result in a difference in A. They can also account for the difference between direct and indirect causation. The problem with manipulability theories is that they are transparently circular. What is it to manipulate A but to cause it? Indeed, when one confronts the details, the manipulation of A must be of a particular type. It must be fully effective at determining A's value but have no direct effect on B except through the action of A.

I favor manipulability accounts because they do not attempt to reduce causation and because they handle causal asymmetry and direct vs. indirect causation perfectly. Using the notation X set= x to mean X is manipulated, or set to equal x, we can define both cause and direct cause as changes in the probability of an effect after manipulation as follows:

X is a *cause* of Y relative to a set of background conditions B just in case for some $x1 \neq x2$, $P(Y \mid X$ *set*$= x1) \neq P(Y \mid X$ *set*$= x2)$.

X is a *direct cause* of Y relative to a set of variables $\mathbf{Z}$ and a set of background conditions B just in case for some $x1 \neq x2$ and some set of values $\mathbf{z}$, $P(Y \mid \mathbf{Z}$ is *set*$= \mathbf{z}, X$ *set*$= x1) \neq P(Y \mid \mathbf{Z}$ is *set*$= \mathbf{z}, X$ *set*$= x2)$.

Manipulation theories take one sort of causation as primitive but make the notion more broadly intelligible insofar as we can imagine what it would be to intervene upon a system and "set" the value of some variable in the system. In many cases we cannot actually perform such a manipulation but can well imagine it. For example, we believe the moon causes the tides, and although we cannot intervene upon the moon much, we can

imagine manipulating the moon's position or eliminating its existence altogether. Our experience as toddlers is one long causal discovery via manipulation – we directly change anything we can get our hands on and
observe what happens next. Breaking stuff is extremely informative
because it is big-time causal discovery. Most, if not all, of our intuitions
about causation, the same intuitions against which we hold philosophical
theories responsible, are extrapolated from these primitive experiences of
manipulation. Nevertheless, if our philosophical goal is to reduce causation to better-understood primitives, the manipulation account is quite
unsatisfactory.

Although this discussion is far too facile, I believe it is ultimately fair.
After two millennia there is still no viable reductive analysis of causation,
and no reason to believe one is forthcoming.

The Computational Problem: Searching for Causal Graphs

The computer forces a totally different perspective on the subject.
Forgetting for a moment what exactly it means to say that one variable X
is a cause of another Y, we must at least formally represent causal structures before we can compute anything about them.[2] We want to represent
causal claims on several levels. On the most general or abstract qualitative
level, we want to represent nothing more than the claim that one variable
is the cause of another, leaving aside all specifics about the strength of the
causal relationship, and so forth. This can easily be done with a class of
formal objects that has been central to computer science almost since its
inception: directed graphs.

A directed graph $G = <V,E>$ is a set of nodes V and a set of edges E,
that is, ordered pairs of nodes. By stipulating that a directed graph represents a causal structure just in case the nodes are variables and an edge is
present from X to Y just in case X is a direct cause of Y relative to V, we
immediately couple causation to computer science.

Two problems come to the fore. One, how does a causal graph, without
any further quantitative elaboration, connect with empirical evidence?
That is, what sorts of evidence does causal structure alone explain? What
set of predictions might one causal graph make and another not, allowing
us to distinguish between them? When are two models empirically indistinguishable? Two, are there efficient methods for searching for the graph
or graphs that explain a given body of evidence? Exactly what assumptions must such techniques rely upon? Can such methods succeed even
when some variables are left unmeasured?

Consider the search problem first. The combinatorics are daunting.
Among only two variables, there are four possible causal arrangements
(figure 2).

---

[2] In this essay I deal only with causation among variables.

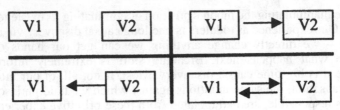

Figure 2. All causal graphs among two variables

For three variables, there are sixty-four possible graphs, and for as few as nine variables there are 4,722,366,482,869,645,213,696 different causal graphs. In general, the total number of graphs among n variables is the number of possible ways each pair can be connected, to the power of the number of pairs of variables:

$$\left(\frac{n(n-1)}{2} \right)_4.$$

Even if we ignore graphs that represent structures with feedback, that is, systems in which one variable is a direct or indirect cause of itself, the number of graphs is still exponential in the number of variables.

If we include the possibility that latent (unmeasured) variables might be common causes of two of our variables, the number of possible ways in which a pair of variables might be connected becomes infinite.

We can get around the infinity problem if we can collapse an infinity of latent-variable models that are equivalent (for example, the lower two models in the left-hand column of figure 3) with respect to the measured variables into one object to search over. How to form such equivalence classes is a subject unto itself, however, which involves appropriately connecting causal graphs in general to empirical evidence. Thus, searching for the "right" causal graph or graphs among those that might govern a small system of variables (ten or so) cannot be done simply by exhaustively visiting each in turn. It is a task that in itself requires serious study.

Causal Graphs and Statistical Evidence

Connecting causal graphs to empirical evidence is not a problem unique to the computational perspective on the subject, but it has a special urgency in this perspective because search cannot even begin until it is solved.

As I said above, on the most general or abstract qualitative level, we want to represent nothing more than the claim that one variable is the cause of another, leaving aside all specifics about the strength of the causal relationship, and so on, and that this is accomplished via causal graphs. The

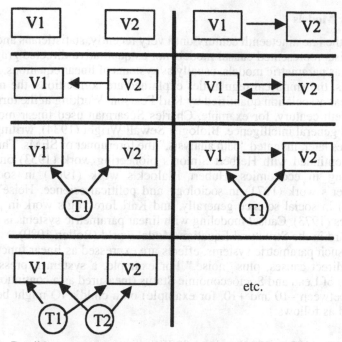

Figure 3. Possible arrangements over two variables, including latent variables

goal, then, is to connect causal graphs to empirical regularities we might test, especially probabilistic or statistical sorts of regularities.

In the 1930s, Sir Ronald Fisher (1935) revolutionized statistical methodology by solving this problem for experimental science. Given a putative cause X and an effect Y, Fisher provided a detailed method for statistically testing whether X is a cause (direct or indirect) of Y. His test involved two pieces. One was instructions on how to assign randomly the value of X for different individuals in the experiment, and the other was instructions for how to compute a "null distribution" against which to compare the outcome of the experiment, that is, the possible outcomes and their expected relative frequencies if X has no effect whatsoever on Y. Although this was an amazing breakthrough, and still constitutes the methodological core of what the FDA requires of studies aimed at establishing the causal effect of some drug or new medical procedure, it requires that one can manipulate (set) the value of the putative cause X. In lots of contexts one cannot achieve this level of control for ethical or practical reasons. For example, in investigating whether HIV is truly the cause of AIDS in humans, we cannot randomly assign some group to "treatment" and infect it with HIV.

The real problem, then, is connecting causal graphs to empirical evidence in *nonexperimental* settings.

Linear Models

From the late nineteenth century until very recently, statisticians and social scientists represented causal models not as qualitative directed graphs but rather as parametric models, usually as systems of linear equations. In such models, the empirical regularities explained are correlations, the measure of linear association quantified by Karl Pearson. Working at the turn of the twentieth century, for example, Charles Spearman used linear models to model general intelligence. Biologist Sewall Wright (1934), writing a few decades later, invented "path analysis," the forerunner of SEMs. This tradition continued with Herbert Simon's pioneering work (1953) on causal ordering in economics, Hubert Blalock's work (1961) in sociology, Costner's work (1971) in sociology and political science, Heise's work (1975) in social science generally, and Karl Joreskog's work in psychometrics (1973). Causal modeling with linear parametric systems is still the standard in the Structural Equation Model world (Bollen 1989).

In such parametric systems, effects are expressed as linear functions of their direct causes, plus "noise." For example, a system expressing the effects of Lead and Socioeconomic Status (measured as a continuous variable between -10 and $+10$, for example) on a child's IQ might be represented as follows:

1. $IQ = 100 - .273*Lead + 1.0*SES + \varepsilon_{IQ}$
2. $Lead = 10 - 2.0*SES + \varepsilon_{lead}$

The linear coefficients represent the dependence of the left-hand-side variable's expectation on the value of nonerror variables on the right-hand side. The noise or error terms ε_{IQ} and ε_{lead} represent "all other unmeasured causes" as well as intrinsic indeterminacy. Typically assumed to

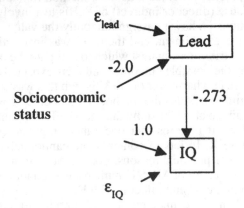

Figure 4. Path diagram

have a normal distribution with mean 0, error terms have no effect on the expectation of the effect, only on its variance. Thus, the model in figure 4 asserts that when Lead and SES both equal 0, IQ averages 100. For every unit increase in Lead exposure, we can expect a decrease of just over a quarter of an IQ point (−.273), and this *dependence* does not depend on the level of the other variables in the equation, for example, Socioeconomic Status.

Representing causation with such systems is appealing, but not unproblematic. For one thing, the world is not always linear. For another, algebraic equations are perfectly symmetric but causation is not. Nothing in the algebraic representation prevents us from transforming an equation like $Y = aX + \varepsilon_y$ to $X = 1/aY + -1/a \, \varepsilon_y$. But in asserting that the equation: $Y = aX + \varepsilon_y$ is a parametric model of the claim: X is a cause of Y, we certainly do not want it entailed that Y is also a cause of X.

Enriching the algebraic representation in order to capture the directionality of causation was crucial. Wright took the first step in the 1930s by associating a path diagram with such systems.

Each arrow in the diagram represents a direct cause, with the linear coefficient representing the strength of the causal relation attached. Thus, it is not enough to simply write down a system of equations; one must also attach a path diagram or, equivalently, designate one form of the equations as canonical. This tradition led to wonderful work from Wright himself on predictions about the "overall effect" of one variable on another, a calculation that depended upon identifying all the causal paths from one variable to another and adding their contribution in a particular way. In the 1950s and 1960s, Simon (1953) and Blalock (1964) took this class of models much farther by deriving "prediction equations" involving *relationships* among the correlations that are entailed solely by the structure of the diagram. Whereas Fisher laid down the first systematic treatment of the epistemology of causation in experimental contexts, Simon's and Blalock's work was the beginning of a systematic epistemology of causation from nonexperimental data, but it does not get the credit it deserves.

In a path diagram like the one in figure 5, where X, Y, and Z are standardized to have mean 0 and variance 1, the correlation between X and Z is equal to the correlation between X and Y times the correlation between

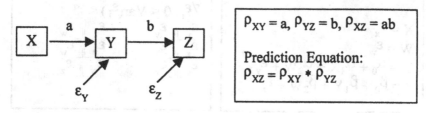

$$\rho_{XY} = a, \ \rho_{YZ} = b, \ \rho_{XZ} = ab$$

Prediction Equation:
$$\rho_{XZ} = \rho_{XY} * \rho_{YZ}$$

Figure 5. Simon-Blalock prediction equations

Y and Z, no matter what value we give to a and b. This constitutes an empirically testable signature of causal structure.

Although the path diagram enriches the representational capabilities of an algebraic system of equations to impose causal direction, it does not obviously encode actual constraints as a result of this directionality.

Writing for a philosophical audience and with no obvious connection to the tradition of Wright, Simon, or Blalock, Dan Hausman (1984) developed an account of causal priority that perfectly filled this gap. Hausman argued that different causes of the same effect could be probabilistically independent, but that different effects of the same cause cannot, thereby giving an account of causal asymmetry in terms of probabilistic independence.

Hausman's insight can immediately be applied to the linear causal model tradition, as I showed in my doctoral thesis. If in an equation like: $Y = aX + \varepsilon_y$ we also insist that the causes of Y, that is, X, and ε_y, are independent, then we induce an asymmetry. The equation itself commutes: that is, $X = 1/aY + -1/a\,\varepsilon_y$, but the statistical constraints can no longer be satisfied – we can no longer keep the "causes" in this form of the equation, that is, Y and ε_y, statistically independent (Scheines 1987).

In the 1980s, the state of the art was to model causal systems as linear systems, with a path diagram that imposed independence constraints on the error terms in the equations. For example, in figure 6 we show the functional interpretation of the causal graph on the left and the statistical constraints (all error terms have nonzero but finite variance and are pairwise independent) imposed on the right.

Whereas Blalock (1964) enumerated linear models of this sort and

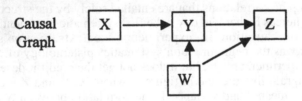

Causal Graph

Functional Relations	*Statistical Constraints*

$X = \varepsilon_x$

$W = \varepsilon_w$

$Y = \alpha_0 + \alpha_1 X + \alpha_2 W + \varepsilon_y$

$Z = \beta_0 + \beta_1 Y + \beta_2 W + \varepsilon_z$

$\forall \varepsilon_i,\ 0 < \mathrm{Var}(\varepsilon_i) < \infty$

$\varepsilon_x \perp\!\!\!\perp \varepsilon_y \qquad \varepsilon_x \perp\!\!\!\perp \varepsilon_w$

$\varepsilon_x \perp\!\!\!\perp \varepsilon_z \qquad \varepsilon_y \perp\!\!\!\perp \varepsilon_w$

$\varepsilon_y \perp\!\!\!\perp \varepsilon_z \qquad \varepsilon_w \perp\!\!\!\perp \varepsilon_z$

Figure 6. Linear causal model

derived prediction equations from them, his work accommodated neither every model nor every possible prediction equation entailed by a model. Glymour, Scheines, Spirtes, and Kelly (1987) extended this work to give a general characterization of when path diagrams without feedback entailed partial correlation constraints (but only up to first order), and they characterized how such models, especially those with latent variables, entailed vanishing tetrad differences. The tradition from Spearman had been elaborated enormously, but it was still stuck in linear land and had not achieved a fully general theory even within that scope. The big leap came from more general representations called Bayes Networks from computer science. It is to them I now turn.

Bayes Networks

In the computer-science community in the early 1980s, artificial-intelligence and robotics researchers faced a daunting problem – how to get an artificial agent like a robot to cope with uncertainty and how to program it to learn from its observations. By representing a robot's state of knowledge as a probability distribution over a set of atomic propositions, we can represent uncertainty, and by "updating" these distributions in response to evidence using Bayes's famous theorem, we can model a rational agent learning. Unfortunately, we cannot just store a probability for each atomic proposition, like "it rained today" or "my lawn is wet"; we must also store joint information, that is, how likely is it that it both rained today and my lawn is wet, as compared to how likely is it that it rained but my lawn is not wet, and so on. The space required to store joint-probability distributions is large, and, because of the joint information, it grows exponentially with the number of atomic propositions. Further, the number of computations required to update a joint-probability distribution, if done naively, is prohibitive. Again, this task grows exponentially with the number of atoms in the algebra.

Fortunately, these are just the sorts of problems computer scientists eat for lunch. Some realized that propositions are often independent. For example, learning that it rained today is quite informative about whether your lawn is wet, but not in the least informative about whether your phone is off the hook. By our taking advantage of such independencies, the space required to store the joint distribution and the number of computations required for updating can be decreased dramatically. Others figured out that directed graphs could encode the independence relationships true of the atomic propositions, and that such graphs could themselves be used to figure out very efficiently exactly how to update the robot's overall knowledge when new evidence came in (Lauritzen and Spiegelhalter 1988).

Figure 7 shows a directed graph functioning as an "Independence Map" (Pearl 1988) that encodes the independence relations over Exposure,

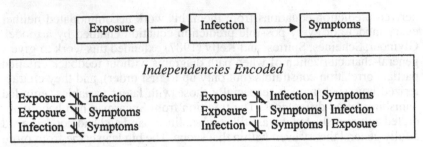

Figure 7. Independence map

Infection, and Symptoms. The map tells us that all pairs are associated, but that Exposure and Symptoms are independent conditional on Infection.

Computer scientists gave us a way to use graphs to represent not just independence relations but full probability distributions that had these independencies. Bayes Networks (Pearl 1988) were among the most popular of these representations, and, like path analysis models, they combine directed graphs with quantitative relationships between variables.

Bayes Networks involve a rule for writing out the joint distribution that follows the structure of the graph. The joint distribution over any set $\mathbf{V} = \{V_1, \ldots, V_n\}$ can be written as a product of conditional distributions:

$$P(\mathbf{V}) = \prod_{i=1}^{n} P(V_i \mid V_1, \ldots, V_{i-1})).$$

In a Bayes Network, however, we condition each variable only on its immediate parents in the directed graph:

$$P(\mathbf{V}) = \prod_{i=1}^{n} P(V_i \mid parents(V_{i-1})).$$

For example, if we take the variables in figure 7, that is, $\mathbf{V} = \{\text{Exp, Inf, Sym}\}$, then we can always encode the joint distribution as a general product of conditionals as we show in the left side of figure 8. Using the directed graph in figure 7, however, we can encode the joint distribution as I show on the right side of figure 8.

This encoding, which implicitly imposes a set of independence relations, makes it possible to represent the joint distribution over a large set of variables very efficiently. The number of parameters needed to store the joint distribution over a set of k Boolean variables is on the order of 2^{k-1}. In the example worked out in figure 7 and figure 8, we can represent the joint probabilities over this trio of variables with seven parameters. By using the independencies, we can reduce the number in this case to five (on the right side of figure 8). In general, an independence map that is a chain (like the one in figure 7) over k Boolean variables reduces the number of parameters needed to store the joint distribution from order 2^{k-1} to order 2k, a drop from exponential to linear complexity, basically the Holy Grail in algorithms.

Although the Bayes Network encoding of a joint distribution implicitly imposes independence constraints, for many purposes computer scientists wanted to make the independencies entailed explicit. Ideally, they wanted a way simply to read the independencies entailed by a Bayes Network encoding off the directed graph, without having to bother with the probability tables at all. Judea Pearl (1988) and some of his students at UCLA solved this problem in the middle of the 1980s, and they called their solution *d-separation*, which stands for dependence separation.

D-separation provides a graphical definition for determining when a set of variables **X** are d-separated from another set **Y** by a set **Z** in a causal graph.[3] Pearl and his student Thomas Verma proved that, in a Bayes Network, if a set of variables **X** are d-separated from another set **Y** by a set **Z** in the directed graph of the network, then **X** _⊥_ **Y** | **Z** in *every* joint distribution representable by that Bayes Network.

General Encoding	Bayes Network Encoding
$P(Exp = yes) = \phi_1$	$P(Exp = yes) = \phi_1$
$P(Exp = no) = 1 - \phi_1$	$P(Exp = no) = 1 - \phi_1$
$P(Inf = yes \mid Exp = yes) = \phi_2$	$P(Inf = yes \mid Exp = yes) = \phi_2$
$P(Inf = no \mid Exp = yes) = 1 - \phi_2$	$P(Inf = no \mid Exp = yes) = 1 - \phi_2$
$P(Inf = yes \mid Exp = no) = \phi_3$	$P(Inf = yes \mid Exp = no) = \phi_3$
$P(Inf = no \mid Exp = no) = 1 - \phi_3$	$P(Inf = no \mid Exp = no) = 1 - \phi_3$
$P(Sym = yes \mid Inf = yes, Exp = yes) = \phi_4$	$P(Sym = yes \mid Inf = yes) = \phi_4$
$P(Sym = no \mid Inf = yes, Exp = yes) = 1 - \phi_4$	$P(Sym = no \mid Inf = yes) = 1 - \phi_4$
$P(Sym = yes \mid Inf = yes, Exp = no) = \phi_5$	$P(Sym = yes \mid Inf = no) = \phi_5$
$P(Sym = no \mid Inf = yes, Exp = no) = 1 - \phi_5$	$P(Sym = no \mid Inf = no) = 1 - \phi_5$
$P(Sym = yes \mid Inf = no, Exp = yes) = \phi_6$	
$P(Sym = no \mid Inf = no, Exp = yes) = 1 - \phi_6$	
$P(Sym = yes \mid Inf = no, Exp = no) = \phi_7$	
$P(Sym = no \mid Inf = no, Exp = no) = 1 - \phi_7$	

Figure 8. Efficient Representation from Bayes network

[3] See the module on d-separation in www.phil.cmu.edu/projects/csr.

Causal Bayes Networks

Toiling away in the land of linear causal models, my group became aware of Bayes Networks and d-separation around 1988. We soon realized that the linear causal models we had been studying were a special kind of Bayes Network, but a Bayes Network just the same. It immediately became apparent that d-separation provided the general link between causal structure and empirical regularity we had been looking for.

What needed to happen, however, is that the edges in the directed graphs associated with Bayes Networks had to be *interpreted as representing direct causation*. Even if one could attach such an interpretation to linear causal models, the question was, why do so generally?

We took two approaches to this question. First, one can generalize the linear causal model framework to what Pearl calls a *functional causal model*. In figure 9, we show the same example as in figure 6, the only difference being that the linear functions in figure 6 are replaced by arbitrary ones in figure 9. So long as the graphs attached to such models are acyclic, that is, have no paths from a variable back to itself, then such models are Bayes Networks and d-separation characterizes the independence relations implied by the causal structure alone.

Second, one can formulate axioms that make explicit the assumptions required to connect causal structure to probabilistic independence in a way equivalent to a Bayes Network. This approach resulted in the Causal Markov Axiom (Spirtes, Glymour, and Scheines 1993, chap. 3), which can be stated as follows

Causal Graph

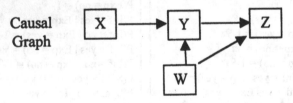

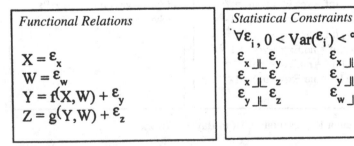

Functional Relations

$$X = \varepsilon_x$$
$$W = \varepsilon_w$$
$$Y = f(X,W) + \varepsilon_y$$
$$Z = g(Y,W) + \varepsilon_z$$

Statistical Constraints

$$\forall \varepsilon_i, \ 0 < \text{Var}(\varepsilon_i) < \infty$$

$$\varepsilon_x \perp\!\!\!\perp \varepsilon_y \qquad \varepsilon_x \perp\!\!\!\perp \varepsilon_w$$
$$\varepsilon_x \perp\!\!\!\perp \varepsilon_z \qquad \varepsilon_y \perp\!\!\!\perp \varepsilon_w$$
$$\varepsilon_y \perp\!\!\!\perp \varepsilon_z \qquad \varepsilon_w \perp\!\!\!\perp \varepsilon_z$$

Figure 9. Functional causal model

A causal graph G over a set of variables **V** satisfies the Causal Markov Axiom just in case in every probability distribution P(**V**) that G can produce, each V ∈ **V** is independent of all other variables in **V** besides V's effects, conditional on V's direct causes.

This axiom is constructed from two intuitions, one from Markov and one from philosophers Hans Reichenbach (1956) and Wes Salmon (1980). In Markov processes, future states are independent of past states given current states. Put another way, variables are independent of their indirect causes given their direct causes. Reichenbach and Salmon discussed how a common cause "screens off" its effects, the upshot being that two variables not directly causally related are independent conditional on all of their common causes.

Causal structures with no feedback that satisfy the Causal Markov Axiom also satisfy d-separation. Interestingly, linear causal models with feedback do not satisfy the Causal Markov Axiom, but they do satisfy d-separation.

By the early 1990s it was clear that the computer-science community and the philosophical community had teamed up to produce a plausible and quite general account of how causal structure alone connected to empirical regularity. What was left was to sketch (1) how to model the use of causal knowledge to predict the effect of interventions, and (2) how to search automatically for causal models from data with the computer.

Causal Prediction: Modeling Manipulations

One clear benefit of having causal knowledge is being able to predict the effect of a manipulation or intervention. For example, smoking and getting lung cancer are positively associated, as are having tar-stained fingers and getting lung cancer. As it is extremely difficult to induce people to stop smoking, perhaps a better public policy to combat lung cancer is to provide people with good tar-solvent soap and advise them to scrub the stains off their fingers every day. How do we know this policy is a joke? We know because we have a causal theory that tells us how the system would *respond* to our tar-solvent-soap intervention.

The soap-intervention seizes all influence over the Tar-stained Fingers variable, and we model this by "x-ing" out the edge from Smokes to Tar-stained Fingers. In the resulting structure, Tar-stained Fingers and Lung Cancer are no longer even associated.

The simple way to model an intervention on a causal graph (see figure 10), which we first articulated coherently in 1989[4] but which was anticipated by Haavelmo as far back as 1943, is to erase all arrows in the causal graph that go into any variable intervened upon. We can model such an

[4] Reprinted as chapter 2 in Glymour and Cooper 1999.

intervention on a Causal Bayes Network by replacing any of the conditional probabilities of the variables intervened upon with the probability distri-bution imposed upon them by the intervention. For example, using the causal graph from figure 7 and the joint distribution encoded as a Bayes Network from the right side of figure 8, I show how we convert the graph and the accompanying distribution to model quantitatively an intervention in figure 11.

In our 1993 work, we describe a more complicated version of this move, but the key insight remains the same. Interventions change the causal structure in a predictable way:

Interventions change the relationship between a variable and its causes but leave intact the relationships between a variable and its effects.

Causal Discovery

Given all this, what causal knowledge can be discovered automatically from nonexperimental data? The answer, as you would expect, depends almost entirely on how much you are willing to assume. The study of exactly what can be discovered and what cannot under what assumptions has become what I call the computational epistemology of causal science.

What sorts of assumptions am I referring to? Assumptions like the

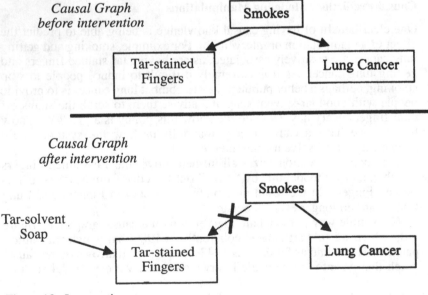

Figure 10. Intervention

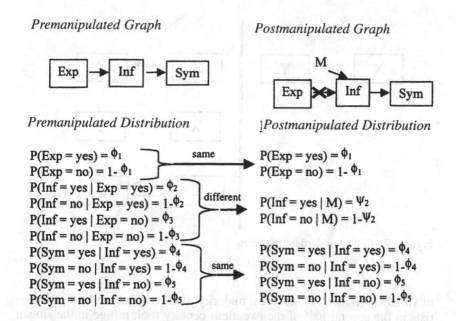

Figure 11. Quantitatively modeling an intervention

Causal Markov Axiom, the assumption that no feedback exists, and the assumption that no latent common causes are active. If one is not even willing to assume the Causal Markov Axiom, or something like it, the game cannot even begin. Even if one is willing to assume that causal structures satisfy it, the first thing that should attract attention is that causal graphs are typically underdetermined by nonexperimental evidence. In figure 12, for example, we show in the left-hand column the graph over just two variables, X and Y, that implies by d-separation that X and Y are independent, and in the right-hand column the three graphs that imply by d-separation that X and Y are associated.

If latent common causes might be acting, the underdetermination is worse. Further, although d-separation characterizes the independencies entailed by just the graph's structure, a Bayes Network may still entail an independence via a very special assignment of values to the parameters that is not entailed by the graph alone, as I discussed for the model of Tax Revenues in figure 1. Thus, a further assumption involves treating all d-separation equivalent models as empirically indistinguishable. We call such an assumption Faithfulness, and it is by no means uncontroversial.[5]

Aware of the underdetermination of causation by association for a pair

---

[5] See, for example, the debate in part 3 in Glymour and Cooper 1999.

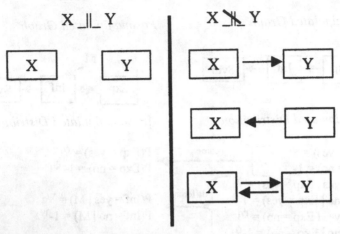

Figure 12. Causal underdetermination

of variables (figure 12), the great majority of statisticians and social scientists in the second half of the twentieth century took refuge in the slogan "correlation does not imply causation," and in doing so virtually annihilated the subject of causal discovery from nonexperimental data. An entire community simply decided that because causal discovery is impossible among a single pair of variables it must be impossible among systems involving more than two variables. The generalization from two to more variables is not only fallacious but even turns out to be the opposite of what is true. The more variables in a system, the more one can discover about what is causing what. The patient indeed lives and breathes!

In a system of three variables $V = \{X, Y, Z\}$, for example, suppose that X and Z are found to be independent, but all other pairs are associated. Assuming the Causal Markov Axiom and Faithfulness, but nothing else, we can conclude that Y is a cause of neither X nor Z. Under similar assumptions, in a system of four variables $V = \{X, Y, Z_1, Z_2\}$, if the independence relations found to hold in the data are $Z_1 \perp\!\!\!\perp Z_2$, $Z_1 \perp\!\!\!\perp Y \mid X$, and $Z_2 \perp\!\!\!\perp Y \mid X$, then in *every* causal graph consistent with these assumptions and these empirical regularities, X is a direct cause of Y relative to V, X and Y have no latent common cause, and X is a cause of neither Z_1 nor of Z_2!

Figure 13 schematizes the work of the computational epistemologist of causal science. Once the set of assumptions is made clear, then the job of the computational epistemologist is to characterize the equivalence structure among the causal graphs deemed possible. That is, under a given set of assumptions, certain *sets* of graphs will be empirically indistinguishable. Given evidence, this equivalence structure makes our causal knowledge

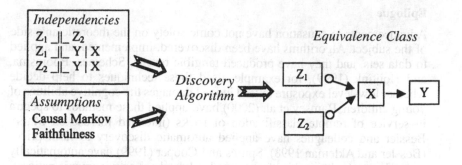

Figure 13. Discovery algorithm

and our causal ignorance precise. A discovery algorithm inputs the evidence and assumptions and outputs this equivalence class. In figure 13, for example, the equivalence class is represented with a graphical object called a Partial Ancestral Graph (Richardson 1996). In this example, the equivalence class contains graphs in which Z_1 is either a direct cause of X or there is a latent common cause of Z_1 and X, similarly for Z_2 and X. But in all members of the class, X is a direct cause of Y.

Besides the Causal Markov Axiom and Faithfulness, here are some of the assumptions studied since the early 1990s:

- D-separation
- Causal Sufficiency (no latent common causes)
- Feedback
- Linearity

Assuming d-separation and Faithfulness, we know how to characterize equivalence over the following classes of models:

1. Causally sufficient, no feedback, linear.
2. Causally sufficient, no feedback, not linear.
3. Causally sufficient, feedback, linear.
4. Not causally sufficient, no feedback, linear.
5. Not causally sufficient, no feedback, not linear.

When and why one should endorse these various assumptions is another topic, but the moral should be clear. What one can discover depends on the assumptions made and the data collected, and the fine-grain structure is as rich and complex for the theory of causation as for any subject I know of.

Epilogue

Advancements in causation have not come solely on the theoretician's side of the subject. Algorithms have been discovered, implemented, and applied to data sets, and they have produced tangible results. Scheines, Boomsma, and Hoijtink (1999), for example, used these techniques to help decide whether low-level exposure to lead indeed damages the cognitive abilities of young children. Ramsey et al. (2000) have applied these methods to spectra in service of remote classification of rocks (intended for use on Mars), Bessler and colleagues have applied automatic discovery to farm prices (Bessler and Akleman 1998), Spirtes and Cooper (1999) have automatically learned causal relationships from a medical database on pneumonia patients, and, most recently, we have begun to apply the techniques to learn about genetic regulatory structure (Spirtes, Glymour, and Scheines 2000a).

The computer has had an incalculably large impact on the theory, and especially the epistemology, of causation. The survey I have given here is no more than the briefest sketch, however. To learn more, I recommend four books that cover most of the field and give references to the parts they do not. The second edition of Spirtes, Glymour, and Scheines (2000) has the most extensive treatment of model equivalence, discovery algorithms, the axiomatization of causal models, and the faithfulness debate. Pearl (2000) gives the best comprehensive treatment of the modern representation of the subject, and is the clearest at distinguishing between observation and manipulation and carrying that distinction all the way through the formalization of the topic. Two edited collections, one by Glymour and Cooper (1999), and one by McKim and Turner (1997), pull together a wide array of important writers, many of whom do not agree with each other and are not afraid to say so. These volumes bring to life the debates that still rage about the subject and make the computational turn in the philosophy of causation accessible.

References

Bessler, D., and D. Akelman. 1998. "Farm Prices, Retail Prices, and Directed Graphs: Results for Pork and Beef." *American Journal of Agricultural Economics* 80: 1144–49.

Blalock, H. 1961. *Causal Inferences in Nonexperimental Research.* Chapel Hill: University of North Carolina Press.

———. 1971. *Causal Models in the Social Sciences.* Chicago: Aldine-Atherton-Atherton.

Bollen, K. 1989. *Structural Equations with Latent Variables.* New York: Wiley.

Costner, H. 1971. "Theory, Deduction, and Rules of Correspondence." In *Causal Models in the Social Sciences*, edited by H. Blalock. Chicago: Aldine-Atherton.

Fisher, R. 1935. *The Design of Experiments*. Reprinted in 1951. Edinburgh: Oliver and Boyd.

Glymour, C., R. Scheines, P. Spirtes, and K. Kelly. 1987. *Discovering Causal Structure*. San Diego: Academic Press.

Glymour, C., and G. Cooper. 1999. *Computation, Causation, and Discovery*. Boston: AAAI Press and MIT Press.

Harary, F., and E. Palmer. 1973. *Graphical Enumeration*. New York: Academic Press.

Hausman, D. 1984. "Causal Priority." *Nous* 18: 261–79.

Haavelmo, T. 1943. "The Statistical Implications of a System of Simultaneous Equations." *Econometrica* 11: 1–12. Reprinted in D. Hendry and M. Morgan, eds., *The Foundations of Econometric Analysis*, 477–90. Cambridge: Cambridge University Press.

Heise, D. 1975. *Causal Analysis*. New York: Wiley.

Joreskog, K. 1973. "A General Method for Estimating a Linear Structural Equation." In *Structural Equation Models in the Social Sciences*, edited by A. Goldberger and O. Duncan. New York: Seminar Press.

Kiiveri, H., and T. Speed. 1982. "Structural Analysis of Multivariate Data: A Review." In *Sociological Methodology*, edited by S. Leinhardt. San Francisco: Jossey-Bass.

———, and J. Carlin. 1984. "Recursive Causal Models." *Journal of the Australian Mathematical Society* 36: 30–52.

Lauritzen, S., and D. Spiegelhalter. 1988. "Local Computations with Probabilities on Graphical Structures and Their Application to Expert Systems [with discussion]." *Journal of the Royal Statistical Society*, ser. B, 50: 157–224.

Lewis, D. 1973. "Causation." *Journal of Philosophy* 70: 556–72.

Mackie, J. 1974. *The Cement of the Universe*. New York: Oxford University Press.

McKim, S., and S. Turner. 1997. *Causality in Crisis?: Statistical Methods and the Search for Causal Knowledge in the Social Sciences*. Notre Dame: University of Notre Dame Press.

Pearl, J. 1988. *Probabilistic Reasoning in Intelligent Systems*. San Mateo: Morgan and Kaufman.

Ramsey, J., P. Gazis, T. Roush, P. Spirtes, and P. Glymour. 2000. "Automated Remote Sensing with Near Infrared Reflectance Spectra: Carbonate Recognition." http://www.phil.cmu.edu/rockspec. Under review at *J. Knowledge Discovery and Data Mining*.

Richardson, T. 1996. "Models of Feedback: Interpretation and Discovery." Ph.D. thesis, Department of Philosophy, Carnegie Mellon University, Pittsburgh.

Reichenbach, H. 1956. *The Direction of Time*. Berkeley: University of California Press.

Rubin, D. 1974. "Estimating Causal Effects of Treatments in

Randomized and Nonrandomized Studies." *Journal of Educational Psychology* 66: 688–701.

Salmon, W. 1980. "Probabilistic Causality." *Pacific Philosophical Quarterly* 61: 50–74.

Scheines, R. 1987. "Causality in the Social Sciences." Doctoral dissertation, Department of History and Philosophy of Science, University of Pittsburgh, Pittsburgh.

―――, A. Boomsma, and H. Hoijtink. 1999. "Bayesian Estimation and Testing of Structural Equation Models." *Psychometrika* 64: 37–52.

Simon, H. 1953. "Causal Ordering and Identifiability." In *Studies in Econometric Methods*, edited by Hood and Koopmans, 49–74. New York: Wiley.

―――. 1954. "Spurious Correlation: A Causal Interpretation." *JASA* 49: 467–79.

Spirtes, P., and G. Cooper. 1997. "An Experiment in Causal Discovery Using a Pneumonia Database." *Proceedings of AI and Statistics* 99.

―――, C. Glymour, and R. Scheines. 1993. *Causation, Prediction, and Search*. Dordrecht: Springer-Verlag.

―――. 2000a. *Causation, Prediction, and Search*. Second edition. Cambridge: MIT Press.

―――. 2000b. "Constructing Bayesian Network Models of Gene Expression Networks from Microarray Data." *Proceedings of the Atlantic Symposium on Computational Biology, Genome Information Systems, and Technology*.

Suppes, P. 1970. *A Probabilistic Theory of Causality*. Amsterdam: North-Holland.

Wermuth, N., and S. Lauritzen. 1983. "Graphical and Recursive Models for Contingency Tables." *Biometrika* 72: 537–52.

Whittaker, J. 1990. *Graphical Models in Applied Multivariate Statistics*. New York: Wiley.

Wright, S. 1934. "The Method of Path Coefficients." *Ann. Math. Stat.* 5: 161–215.

12

PHILOSOPHY FOR COMPUTERS:
SOME EXPLORATIONS IN PHILOSOPHICAL MODELING

PATRICK GRIM

Thought experiments and conceptual analogies have played a major role in philosophical thinking from Plato's cave to Rawl's original position. Thought experiments and conceptual analogies can be thought of as models, and in that regard philosophical modeling is nothing new. The history of logic, in fact, can be read as a history of one particular type of philosophical modeling – an attempt to construct rule-governed syntactic models adequate to track valid patterns of inference.

In this sense, modeling has long been a major element of the philosopher's toolkit. What is new is the computational power now readily available for modeling in general. Computers, including the personal computer now idling on your desk, are astoundingly powerful modeling machines. Even the least expensive contemporary computer allows the creation, manipulation, and experimental modification of models with a level of intricacy and with a computational speed and depth undreamt of only twenty years ago. Computational modeling has become increasingly important in both the physical and social sciences, particularly in physics, theoretical biology, sociology, and economics. Given the strong role that conceptual modeling has always played in philosophy, it seems inevitable that the computer will become a major tool in philosophical work as well.

Over the past dozen years or so, the informal Group for Logic and Formal Semantics in the philosophy department at SUNY Stony Brook has been exploring a range of philosophical questions in terms of computer models. Undergraduates, graduate students, visitors, and several faculty members have all played major roles in the work of the group, using the resources of a quite standard bank of teaching computers in the Logic Lab. A representative sample of our work appears in *The Philosophical Computer* (Grim, Mar, and St. Denis 1998), with small bits featured several times in *Scientific American* (Stewart 1993, 2000). Our current work finds a niche not only in philosophy but also in major journals in theoretical biology, linguistics, decision theory, engineering, and computer graphics. Computational modeling is a new frontier across a range of disciplines, and one of the exciting aspects of this frontier is that interdisciplinary work tends to be the general rule rather than the exception.

What I want to offer here is an introduction to just a handful of our explorations in philosophical modeling: explorations regarding (1) the potential emergence of cooperation in a society of egoists, (2) self-reference and paradox in fuzzy logic, (3) a fractal approach to formal systems, and (4) on-going explorations with models for the emergence of communication. We are but one of a number of teams working from a variety of approaches on a range of questions in philosophical computation, and indeed many other teams have a more formal organizational structure, better funding, and a tighter concentration. I have attempted a wider survey of contemporary philosophical modeling elsewhere (Grim 2002). Here I want to focus on a few examples of directions that our own work has taken. In each example what I am after is a brief introduction to the basic ideas that guided our work in a particular area, rather than a detailed summary of methods and results. A paper trail of further references is provided for those who want to search out the details or who may want to carry the techniques further in research of their own.

The Growth of Cooperation in a Hobbesian World

What Hobbes imagines is an initial condition without government and in which all are out for themselves alone: an anarchic society of self-serving egoists. This is Hobbes's "condition of Warre of every one against every one," in which "every man is Enemy to every man" and life as a result is "solitary, poore, nasty, brutish, and short" (Hobbes 1997).

How might social cooperation emerge in a society of egoists? This is Hobbes's central question, and one he attempts to answer in terms of two "general rules of Reason." As there can be no security in a state of war, he argues, it will be clear to all rational agents "that every man ought to endeavor peace, as farre as he has hope of obtaining it; and when he cannot obtain it, that he may seek, and use, all helps, and advantages of Warre." From this Hobbes claims to derive a second rational principle: "That a man be willing, when others are so too . . . to lay down this right to all things, and be contented with so much liberty against other men, as he would allow other men against himself" (Hobbes 1997).

Building on earlier work in theoretical biology and sociology, we have developed a range of Hobbesian models in the hope of throwing further light on the emergence of cooperation. The basic question is the same: How might social cooperation emerge within a society of self-serving egoists? Our models, however, embed contemporary game theory in the spatialization of cellular automata. The picture of emerging cooperation that appears using these model-theoretic tools, interestingly enough, seems to parallel Hobbes's second principle.

The most studied model of social interaction in game theory is undoubtedly the Prisoner's Dilemma. Here we envisage two players who must simultaneously make a move, choosing either to 'cooperate' with the other player or to 'defect' against him. The standard matrix for the Prisoner's

Dilemma lays down how much each player will gain or lose on a given move, depending on the mutual pattern of cooperation and defection. Here player B's gains in each contingency are shown to the left of the comma, player A's gains to the right:

		Player A	
		Cooperate	Defect
Player B	Cooperate	3, 3	0, 5
	Defect	5, 0	1, 1

If both players cooperate on a single move, each gets three points. If both defect, each gets only one point. But if one player defects and the other cooperates, the defector gets a full five points and the cooperator gets nothing.

The philosophical force of the Prisoner's Dilemma, of course, is the stark contrast it paints between egoistic rationality and collective benefit. If player B chooses to cooperate, it is to player A's advantage to defect: A then gets five points in place of a mere three for mutual cooperation. If player B chooses to defect, it is still in player A's interest to defect: in that case A salvages at least the one point gained from mutual defection instead of the zero A would get from cooperation in the face of defection. Whatever Player B chooses to do, then, the rational thing for A to do on a single round is to defect: defection is said to be *dominant* in a single round of the Prisoner's Dilemma.

The situation is symmetrical. All information regarding payoffs is common knowledge, and what is rational for one player is rational for the other. Two rational players in a single turn of the Prisoner's Dilemma can thus be expected to defect against each other. But economic rationality on each side can then be expected to result in a payoff situation that is clearly worse than mutual cooperation for each player and that is collectively the least desirable outcome of all. The rational attempt to maximize gains on each side seems predictably, inevitably, to fail.

The Prisoner's Dilemma becomes both more interesting and more realistic as a model for biological and social interaction when it is made open-ended: when players are envisaged as engaging in repeated games, never knowing when or whether they might meet again. In this Iterated Prisoner's Dilemma strategies for play over time emerge, including for example a vicious strategy of universal defection – 'All Defect', or AllD. 'Cooperate Then All Defect', or C-then-AllD, on the other hand, offers initial cooperation as a come-on but follows through with vicious and unremitting defection. Of particular interest in the Iterated Prisoner's Dilemma is the strategy 'Tit for Tat', or TFT, which cooperates to begin with and thereafter simply repeats its opponent's play from the previous round. If you cooperate with TFT, it will return the cooperation on the following round. If you defect against TFT, it will react with a reciprocal defection.

In 1980 Robert Axelrod solicited strategies for a first computer tournament in the Iterated Prisoner's Dilemma, in which submitted strategies were played round-robin against all other competitors, themselves, and a strategy that played at random. Despite the sophistication of many submitted entries, the strategy that emerged as the winner in Axelrod's first tournament was simple Tit for Tat. The winner of that first competition was announced, new strategies were solicited, and – surprisingly – TFT emerged as the winner again (Axelrod 1980, 1984). With William Hamilton, Axelrod then switched to a biological model based on replicator dynamics, in which winners in a given generation had a reproductive advantage resulting in a greater proportion of the population at the next stage (Axelrod and Hamilton 1981; Axelrod 1984). Here again the winner was Tit for Tat.

In our Hobbesian models we add a spatial dimension to the Iterated Prisoner's Dilemma, embedding our Hobbesian individuals as cells in the cellular automata array shown in figure 1.

Unlike previous models, an individual in this Spatialized Prisoner's Dilemma is not pitted against all other players in the field. In the style of John Horton Conway's Game of Life, competitive action is local rather than global.[1] Each cell plays only its immediate neighbors in a succession of games, then totals its points and compares its score with that of its immediate neighbors. If a cell has a neighbor that has been more successful – if a neighboring cell has a higher total score – the cell changes its strategy to match the neighbor.

Suppose, then, we start with a randomized array of simple strategies. In a spatialized model, the strategies that prove more successful can be expected to spread progressively so as to occupy the entire field. That is indeed what happens. Figure 2 shows chosen steps in an evolving array from a randomization of eight simple strategies. 'All Defect' and 'Cooperate Then All Defect' show early growth, but the eventual winner is Tit for Tat.

These explorations using the Spatialized Prisoner's Dilemma fully support earlier work on the robustness of TFT. This history of modeling also seems to vindicate Hobbes's general conclusions: TFT quite effectively instantiates Hobbes's second "general Rule, found out by Reason" that "Whatsoever you require that others should do to you, that do ye to them."

---

[1] See Berlekamp, Conway, and Guy 1982. A popular introduction to the Game of Life is Poundstone 1985. Many interactive programs for the Game of Life are available online.

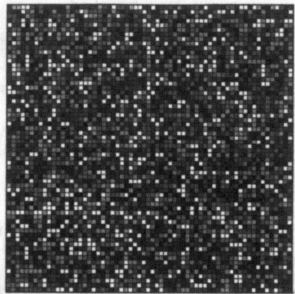

Figure 1. The initial configuration of the Spatialized Prisoner's Dilemma: randomized strategies in a wrap-around array.

The tradition of modeling introduced by Axelrod and Hamilton has been taken one step farther in the work of Martin Nowak and Karl Sigmund (Nowak and Sigmund 1990, 1992; Sigmund 1993). Nowak and Sigmund add a further element of realism into the Axelrod model by introducing stochastic imperfection. The real world is not a world of perfect information and perfect execution, after all. We may not always perceive defections and cooperations for what they are, and we may not always cooperate successfully when we attempt to do so. What happens if we add stochastic grit, modeling an imperfect world by making all of the strategies probabilistically imperfect?

What Nowak and Sigmund found was that in an imperfect world Tit for Tat (or its closest stochastic approximation) is no longer the winner. What wins in an imperfect world is a *more* generous variation called 'Generous Tit for Tat', which does not return every defection with defection but *forgives* defection with a probability of 1/3. A stochastically imperfect world, in other words, seems to increase the role for cooperation. Of particular interest in Nowak and Sigmund's model is that it shows a clear step-by-step increase of generosity in an imperfect world. Up to the appearance of Generous Tit for Tat, dominance by a particular level of generosity is required in order to make way for the emergence of a more generous strategy still (figure 3).

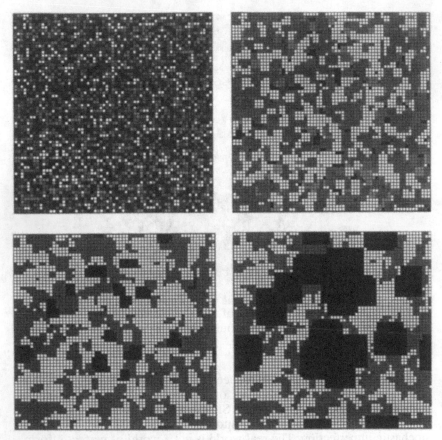

Figure 2. Progressive conquest by Tit for Tat, shown in black, in an array of eight simple strategies. TFT eventually conquers the entire field.

We have added spatialization as a further element of realism in Nowak and Sigmund's stochastic model. Here our initial array is composed of individuals randomized across a large range of probabilistic strategies in the Prisoner's Dilemma, once again competing only locally and imitating the strategies of their most successful neighbors.

With some important qualifications, the strategy that emerges triumphant in this spatialized and stochastic model is one even *more* generous than Nowak and Sigmund's Generous Tit for Tat. Within a spatialized and stochastic Hobbesian world, the strategy with the highest score against itself and impervious to invasion from small clusters of other strategies turns out to be 'Bending Over Backwards', which forgives defection against it with a surprising probability of 2/3 – twice the rate of

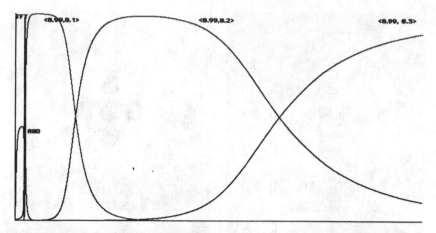

Figure 3. The Nowak and Sigmund result: evolution toward Generous Tit for Tat in a world of imperfect information. The first number of <0.99, 0.3> indicates the probability of returning cooperation with cooperation; the second indicates the probability of returning defection with cooperation.

even Generous Tit for Tat (Grim 1995, 1996; Grim, Mar, and St. Denis 1998). With each step of increasing realism in this tradition of Hobbesian modeling, in other words, we see the emergence of an increased level of cooperation.

Another lesson within this model is that relations between spatialized and stochastic strategies can become intricate, complex, and extremely sensitive to particular configurations. Figure 4, for example, shows stages in the evolution of an array of strategy <0.99999999, 0.33333333> – close to Nowak and Sigmund's Generous Tit for Tat – invaded by a single sixteen-square block of the extremely cooperative <0.99999999, 0.99999999>.

The Dynamics of Paradox

The familiar Liar's paradox is a sentence that asserts its own falsehood:

> The boxed sentence is false.

When people first confronted with such a sentence are pressed to say whether it is true or false, their thinking often oscillates:

"Let's see. Is the sentence true or false? Well, maybe it's true. Oops. If

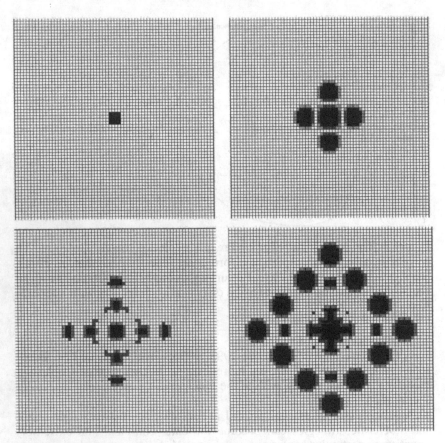

Figure 4. A field of strategy <0.9999999, 0.3333333> invaded by a sixteen-square block of <0.9999999,0.9999999>, shown at intervals of four generations. Vertical lines indicate cells that have been invaded in the last generation. Gray shading indicates those that have reverted to the original strategy in the last generation.

what it says is true, it must be false. So if it's true, it must be false. But then if it's false, it must be true, because it says it's false and that's exactly what it is. So if it's false, it must be true. But then if it's true, it's false, but then it must be true, and so it must be false . . . ."

This kind of informal thinking about the Liar exhibits a clear and simple dynamics: a supposition of 'true' forces us to 'false', the supposition of 'false' forces us back to 'true', the supposition of 'true' forces us back to 'false', and so forth. If one graphed that series of revised reactions, it would look like the oscillation between True and False that appears in figure 5.

What even this simple case suggests is the possibility of using iterated

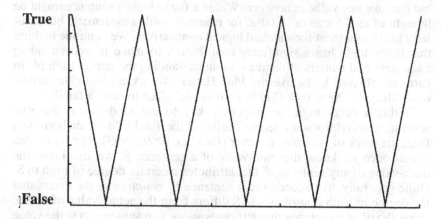

Figure 5. The oscillation of the Classical Liar.

functions to model the phenomena of paradoxical self-reference. Given an initial estimate of its truth or falsity, a sentence that self-referentially attributes a certain truth-value may force one to reevaluate and revise that initial estimate. Given a revised estimate, the sentence may force one to revise the estimate again. Given that further estimate. . . .

The general idea of functional iteration is that of feedback. We input some initial value into a function and obtain an output. That output is then recycled as a new input, giving us a new output. That output again becomes an input, and the cycle of iteration continues.

A standard way of representing progressive values $x_1, x_2, x_3, \ldots$ for an iterated function f is to characterize the value of any x_{n+1} in terms of the application of the function to its predecessor x_n:

$$x_{n+1} = f(x_n)$$

The central idea in our models for paradox is to map truth-attributions in terms of basic functions and to capture self-reference using the idea of iteration. Because the subject of nonlinear dynamics, or chaos theory, is precisely the behavior of iterated functional sequences, this basic idea allows us to apply a range of modeling tools from chaos theory to the philosophical phenomena of self-referential paradox.

Our primary focus has been on infinite-valued or fuzzy logics, in which sentences are not restricted to the two classical values of 'true' or 'false'

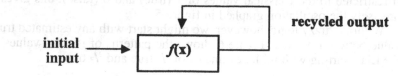

but can take any value in between. Within a fuzzy logic a sentence might be thought of as .75 true or .3 false, for example, with a systematic but similarly fuzzy treatment for standard logical connectives. We continue to think that fuzzy logic has a significant contribution to make in understanding vagueness and matters of degree – in understanding the partial truth of 'In cartoons, Russell looks like the Mad Hatter', for example, or the almost total falsity of 'New York City is a wonderful place to raise a family'.

Within a fuzzy logic, an important key to our modeling is the Vvp schema, a straightforward generalization of the Tarski T-schema borrowed from the work of Nicholas Rescher (Rescher 1969, 1993). The basic idea is that once we know the truth-value of a sentence S, we also know the truth-value of any sentence S' that attributes a certain degree of truth to S. Quite generally, that second-order sentence S' is *un*true to the extent that the degree of truth it attributes to S differs from the actual value of S. The value /V**vp**/ of a sentence that attributes value **v** to sentence **p** is the value of full truth minus the degree of error, gauged as the absolute difference between the attributed value **v** and the real value /**p**/:

$$/V\mathbf{vp}/ = \mathbf{1} - Abs(\mathbf{v} - /\mathbf{p}/)$$

full truth   error of attribution

We use the Vvp schema to model truth attribution, including fuzzy truth-attribution, and use iteration to model self-reference. With these simple tools we can graph and study the semantical dynamics of a range of self-referential and quasi-paradoxical sentences. It should be emphasized, perhaps, that our attempt here is not to *solve* the paradoxes. Philosophical attempts at 'solution' over the past two thousand years have been far from successful; what we are after is simply a deeper understanding of the semantic dynamics involved.

With these basic tools, the Liar can be seen as self-attributing a truth-value of 0. For any estimate /L/ of the truth-value for the Liar, we shall be forced to a revised estimate /VfL/:

$$/VfL/ = 1 - abs(0 - /L/)$$

Where we start with an estimate x_0 for the truth of the Liar sentence, then, we shall be forced to a series of revised estimates $x_1, x_2, x_3, \ldots$ governed by the iterative formula

$$x_{n+1} = 1 - abs(0 - x_n)$$

If restricted to the Classical values of 1 (true) and 0 (false), this gives us precisely the oscillation graphed in figure 5.

Within a fuzzy logic, however, we might start with any estimated truth-value between 0 and 1. Figure 6 shows the patterns of iterated values for the Liar starting with initial estimates of $1/4$ true and $2/3$ true.

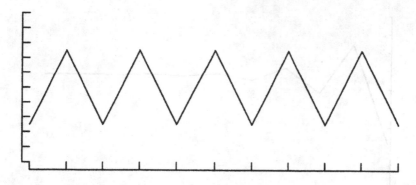

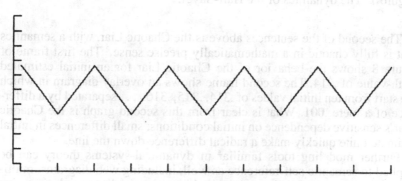

Figure. 6. The dynamics of the Liar in a fuzzy logic with initial values of
¹/₄ and ²/₃.

Once we have the general tools for modeling self-reference in fuzzy
logic, however, there is no reason to restrict our attention to the simple
Liar. We can now consider the dynamics of sentences that self-attribute a
particular fuzzy value, for example, or sentences that self-attribute truth as
a function of our present estimate:

This sentence is ¹/₄ true.

This sentence is as true as you think it is false.

The first of these sentences, which we call the Half-Sayer, has a semantic dynamics that exhibits a fixed-point attractor: starting from any initial
estimate for the Half-Sayer, we are forced to a series of revised estimates
that converges on the value of 2/3. Figure 7 shows the dynamics of the
Half-Sayer for an initial estimated truth-value of .314.

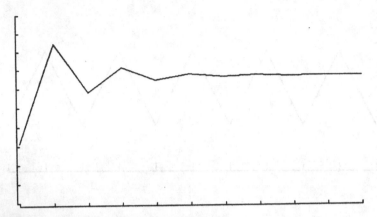

Figure 7. The dynamics of the Half-Sayer.

The second of the sentences above is the Chaotic Liar, with a semantics that is fully chaotic in a mathematically precise sense.[2] The first frame of figure 8 shows the behavior of the Chaotic Liar for an initial estimated truth-value of .314. The second frame shows an overlay diagram in which we start from ten initial values of .314, .315, .316 . . . , separated by a difference of a mere .001. What is clear from this second graph is the Chaotic Liar's sensitive dependence on initial conditions: small differences in initial estimate quite quickly make a radical difference down the line.

Further modeling tools familiar in dynamical-systems theory can be applied to semantic self-reference as well, including web diagrams, multi-dimensional graphs in phase space, and escape-time diagrams.[3]

Beyond single self-referential sentences lie mutually referential pairs of sentences like sentences X and Y in the Chaotic Dualist:

X: X is as true as Y.

Y: Y is as true as X is false.

The strange attractor formed by revised values for X and Y in the Chaotic Dualist is shown in figure 9.

Figure 10 illustrates how this kind of modeling can hint at deep results even when we cannot yet claim to understand them. The figure is an escape-time diagram, which graphs semantic information concerning pairs of initial values for X and Y: in particular, how many iterations are required before a set of revised values takes us beyond a certain distance from the origin (the familiar images of the Mandelbrot set are escape-time

---

[2] See for example Devaney 1989.

[3] A full outline to these techniques is beyond the brief introduction I am attempting here, but details can be found in Mar and Grim 1991 and Grim, Mar, and St. Denis 1998, ch. 1.

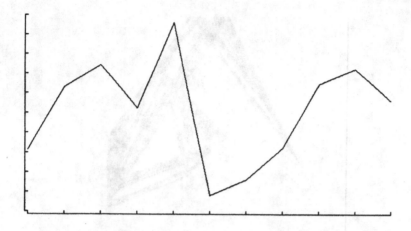

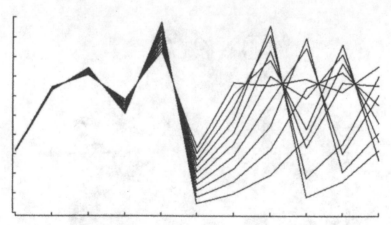

Figure 8. The dynamics of the Chaotic Liar, shown for an initial value .314 (top) and for nine additional values at .001 intervals (bottom).

diagrams as well). The lines of figure 10 track those initial points at which the number of iterations change. Fuzzy truth-values between 0 and 1 are in fact contained in a central region of the diagram; what the pattern as a whole shows us is information for iterated pairs of values were we to think of truth and falsity as going even beyond a [0, 1] interval.

What these explorations promise, we think, is a deeper understanding of the semantic dynamics of a wide range of patterns of self-reference. What

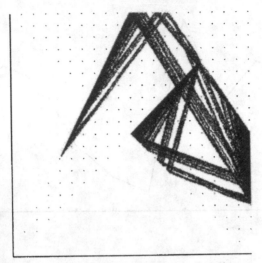

Figure 9. The strange attractor of the Chaotic Dualist.

Figure 10. An escape-time diagram for the Chaotic Dualist for values from
−1.4 to +2.4.

they have already established quite dramatically is the existence of genuinely chaotic phenomena within the semantics of self-referential fuzzy logic.

We have used similar tools to track patterns of tangled reference in the epistemic realm. As epistemic agents, we receive a variety of messages, all claiming to be genuine information, from a variety of sources. Some of the messages received, moreover, offer information or purported information *about* the reliability of other sources of information. It is quite common for a scientific experiment to indicate that an important variable was not in fact held constant in an earlier test, for example; if one of our meters is right, another of our meters must only be roughly and intermittently reliable; two of our friends may be unanimous in telling us that a third friend is not to be trusted. Although it is epistemic reliability that is at issue here, patterns of mutual reference can produce many of the same phenomena documented above – epistemic fixed points, repellor points, and chaos. We have also explored means of reacting to epistemic tangles, however, including models for means of managing, controlling, or minimizing epistemic chaos (Grim, Mar, and St. Denis 1998, ch. 2).

Our modeling work here has also had less-direct consequences, including a theorem in the tradition of Gödel and Tarski on the formal undecidability of chaos: that there is no general algorithm for deciding whether an arbitrary expression of a system adequate for number theory defines a function chaotic on a given interval (Mar and Grim 1991; Grim, Mar, and St. Denis 1998, ch. 1).

Fractal Images of Formal Systems

One of the promises of new modeling techniques is the ability to think of familiar phenomena in new and unfamiliar ways – ways that suggest alternative understandings of the basic phenomena and alternative ways to analyze it. One direction our modeling explorations have taken is an attempt to capture in a visual and immediate way some of the otherwise abstract features of familiar formal systems.

Lukasiewicz, best known for early work in multivalued logics, characterizes his intuitive feelings for logic in terms of independent and unchangeable logical objects:

> Whenever I am occupied even with the tiniest logistical problem, e.g. trying to find the shortest axiom of the implicational calculus, I have the impression that I am confronted with a mighty construction, of indescribable complexity and immeasurable rigidity. This construction has the effect upon me of a concrete tangible object, fashioned from the hardest of materials, a hundred times stronger than concrete and steel. I cannot change anything in it; by intense labour I merely find in it ever new details, and attain unshakeable and eternal truths. (Lukasiewicz 1961, 1970, 22)

We can construct simple Lukasiewiczian objects for the connectives of standard propositional calculus as follows. As any beginning logic student knows, a sentence letter **p** is classically thought of as having two possible values: true or false. It is in terms of these that we draw a simple truth table showing corresponding values for 'not **p**': if **p** happens to be true, 'not **p**' must be false; if **p** happens to be false, 'not **p**' must be true:

$$\textbf{p} \ \textbf{\textasciitilde p}$$
$$\text{T} \ \text{F}$$
$$\text{F} \ \text{T}$$

What we have drawn for **p** and **~p** are two two-line truth tables. But of course these are not the only two-line combinations possible. We expand this to four possibilities if we add the truth tables appropriate for tautologies (always true) and contradictions (always false):

$$\bot \quad \textbf{p} \quad \textbf{\textasciitilde p} \quad \top$$
$$\text{F} \quad \text{T} \quad \text{F} \quad \text{T}$$
$$\text{F} \quad \text{F} \quad \text{T} \quad \text{T}$$

Now consider the possibility of making this information more visual by assigning each of these truth-table columns a particular color, or a contrasting shade of gray:

$$\bot \quad \textbf{p} \quad \textbf{\textasciitilde p} \quad \top$$
$$\text{F} \quad \text{T} \quad \text{F} \quad \text{T}$$
$$\text{F} \quad \text{F} \quad \text{T} \quad \text{T}$$

With these colors we can paint simple portraits of the classical connectives. The top image in figure 11 is a portrait of conjunction: the value colors on its axes combine in conjunction to give the values at points of intersection. The conjunction of black with black in the upper-left corner, for example, gives us black, indicating that the conjunction of two contradictions is a contradiction as well.

The bottom image in figure 11 is a similar portrait for disjunction. Seen together, it is clear that the images have a certain symmetry: the symmetry standardly captured by speaking of disjunction and conjunction as dual operators.[4]

---

[4] On duals see for example Quine 1959.

Figure 11. Color portraits for AND (top) and OR (bottom).

These first images portray truth-table columns as colors, and they are limited to a logic with a single sentence **p**. We can also portray truth-table columns in terms of binary encodings converted to heights (St. Denis and Grim 1997; Grim, Mar, and St. Denis 1998, ch. 3). In a richer logic with two sentential values the resulting Lukasiewiczian value solids take the form shown in figure 12. Progressive additions of sentential values simply deepen the fractal character of these portraits.

Concrete models of this sort both confirm well-known features of familiar formal systems and draw attention to aspects that are as yet little understood. The fact that conjunction and disjunction are duals is exhibited here in the fact that the value solid for AND could be inverted so as to fit perfectly into the spaces of the value solid for OR. The fact that the value solid for material implication is simply a rotation of that for disjunction reflects a geometrical property of negation and the equivalence of $p \rightarrow q$ and $\sim p \vee q$.

One of the features that comes as a surprise in this modeling is the

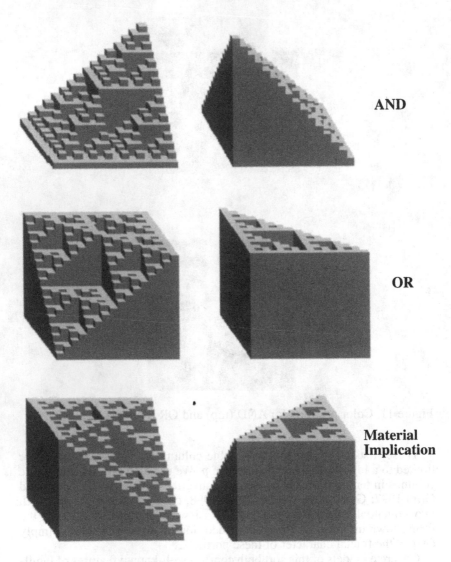

AND

OR

Material Implication

Figure 12. Łukasiewiczian value solids for classical connectives.

appearance of familiar fractals. An image that reappears across solids for different connectives, persistent in increasingly complex languages, is a familiar fractal known as the Sierpinski gasket. For a language with any number of sentence letters, for example, those combinatory values for NAND that turn out to be tautologies form a visible Sierpinski gasket.

This strange fact has led us to another: that a tautology map of this sort can be dynamically generated from a single seed using a simple set of cellular automata rules. In the first frame of figure 13 we start with a single darkened cell in the lower-right corner. Thinking of the top of the grid as 'north' and the bottom as 'south', we follow a simple rule:

A cell will be darkened on the next round if and only if exactly one southeastern neighbor is darkened on the current round.

The result step by step is the progressive generation of a Sierpinski gasket, adequate for portraying those combinatory values for Nand that result in tautologies. Were one able to use a physical instantiation of value solids for real computational purposes, it might be possible to employ a physical parallel of this dynamic process in order to 'grow' usable logic gates (fig. 14).[5]

Our hope is that in the long run modeling approaches of this sort can

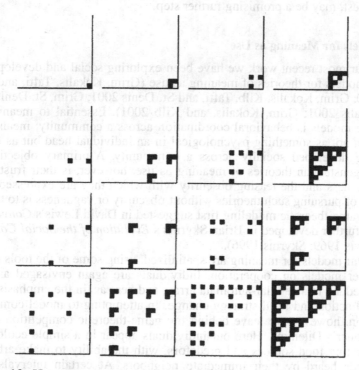

Figure 13. How to grow 'computronium': the dynamic generation of a Sierpinski tautology gasket.

[5] For further elaboration on these speculations concerning 'computronium' see St. Denis and Grim 1997 and Grim, Mar, and St. Denis 1998, ch. 3.

Figure 14. Random-walk migration of a food source or predator across an array of different strategies.

offer a visual glimpse of logical systems as infinite wholes in ways that would suggest genuinely new results. Number-theoretical analysis of logical systems forms a familiar and powerful part of the work of Gödel and others. Analysis in terms of geometry and fractal geometry, we want to suggest, may be a promising further step.

Models for Meaning as Use

In our most recent work we have been exploring social and developmental models for theories of meaning as use (Grim, Kokalis, Tafti, and Kilb 2000; Grim, Kokalis, Kilb, Tafti, and St. Denis 2001; Grim, St. Denis, and Kokalis 2001; Grim, Kokalis, and Kilb 2001). Essential to meaning in these models is behavioral coordination across a community: meaning is taken not as something psychological in an individual head but as something developed socially across a community. A primary objection to Wittgensteinian theories of meaning as use, however, is their frustrating vagueness and the teasing obscurity with which they are expressed. One way of pursuing such theories without obscurity or vagueness is to turn to the game-theoretic modeling first suggested in David Lewis's *Convention* and further developed in Brian Skyrms's *Evolution of the Social Contract* (Lewis 1969; Skyrms 1996).

Our models for meaning are spatialized, using some of the tools of our earlier models on cooperation. Individuals are again envisaged as cells embedded in a cellular-automata array, and here again the emphasis is on local action and local strategy change. In attempting to model communication, however, we leave behind the game-theoretic competition of the Prisoner's Dilemma. Here our individuals appear in a simple ecology of migrating food sources and predators, with the ability to make arbitrary sounds heard by their immediate neighbors. At certain intervals cells imitate the behavior patterns of more successful neighbors. What we want to know in building these models is whether conditions this simple can suffice for the emergence of simple patterns of signaling.

Like individual corals in a coral reef, the individuals in our array do not

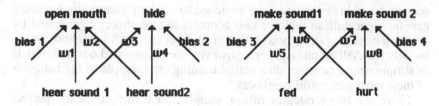

Figure 15. Perceptron architecture.

move. What do move are (1) food sources and (2) predators, each of which wanders cell by cell in a random walk across the array (fig. 15).

Each of our individuals has a small behavioral repertoire. It can open its mouth, hide, or linger in a neutral state, neither opening its mouth nor hiding. It can make a sound 1 or a sound 2 heard by immediate neighbors. Each individual also has a limited range of perception. It knows when it is fed – when its mouth is open and a food source lands on it – and when it is hurt – when not a predator lands on it and it is not hiding. It can hear and distinguish sounds made by immediate neighbors; it knows when a neighbor has just made sound 1, for example, or when a neighbor has just made sound 2.

The behavior of each individual is dictated by a simple strategy. An individual's strategy may dictate that it never opens its mouth, for example, that it opens its mouth only when it hears sound 1, or that it hides only when it hears sound 2. Its strategy may also dictate that it never makes any sounds, or makes sound 2 only when it is fed, or sound 1 only when it is hurt. Calculating all possible behaviors on all possible inputs, we are dealing with an enormous number of possible strategies. But when we start an array we start with a simple randomization across the sample space of those strategies.[6]

When an individual opens its mouth and food is on it, it gains a point for 'feeding'. When a predator hits an individual that is not hiding, that individual is 'hurt' and loses a point. But both opening one's mouth and hiding exact a small energy cost, as does making any sound. After one hundred rounds of gains and losses from feeding and predation – a 'century' – our individuals are given the possibility of strategy change. Each individual surveys the local neighborhood of other cells touching it in order to see if any has garnered more points over the preceding century. If so, it modifies its strategy in the direction of that of its most successful neighbor. In our initial investigations, we used strategy change by pure imitation, precisely as in the Hobbesian models described above. In a second set of investigations, we switched to strategy change by localized genetic algorithm, in which a cell adopts a hybrid strategy generated by crossing its earlier strategy with that of its most successful neighbor. In the models we shall emphasize here, however, we instantiated our individuals as simple neural nets that do a partial training on a sample of the behavior of their most successful neighbors.

There are two strategies in our sample space that count as 'perfect communicators'. Consider, for example, a community of individuals who all share the following strategy:

> They make sound 1 when they are successfully fed and open their mouths when they hear a neighbor make sound 1.

> They make sound 2 when they are hurt and hide when they hear a neighbor make sound 2.

Our food sources migrate from cell to cell, never being consumed. If an individual within a community of perfect communicators feeds – if its mouth is open when a food source lands on it – it will make sound 1 heard by its immediate neighbors. Because its neighbors are also perfect communicators, they will open their mouths in response to the sound, and the migrating food source will land on one of them. *That* individual will then successfully feed, again making a sound that will cause its neighbors to open their mouths. The result is a chain reaction of successful feeding as the food travels across a community of perfect communicators.[7] It should be clear, however, that this advantage demands a full community of perfect communicators. For a lone individual a strategy of 'perfect communication' affords no advantage at all; given the energy cost for sounding and opening its mouth, such a strategy in fact carries a heavy penalty. It should also be clear that there are two flavors of perfect communicators: those that use sound 1 for feeding and sound 2 for predation, and those that use sound 2 for feeding and sound 1 for predation.

Suppose we start, then, with a fully randomized array of individuals with different strategies. Over the course of a hundred turns, our neural-net individuals collect points from gains and losses. At this point we have them survey their neighbors to see if anyone is doing better – whether there is an immediate neighbor that has racked up more points. If so, we have our neural nets do a partial training in that direction. If not, the individual

---

[7] The dynamics of predation is similar but not identical. If a perfect communicator is hurt – is hit by a predator and is not hiding – it will made sound 2. Its neighbors, also perfect communicators, will hide and thus avoid a predator hit. But because the predator does not find a victim on that round, no alarm will be sounded, and thus it will be able to find a victim on the following round. What a strategy of perfect communication avoids is merely predation on every second round. Because of the different dynamics of feeding and predation, we standardly use 100 food items and 200 predators.

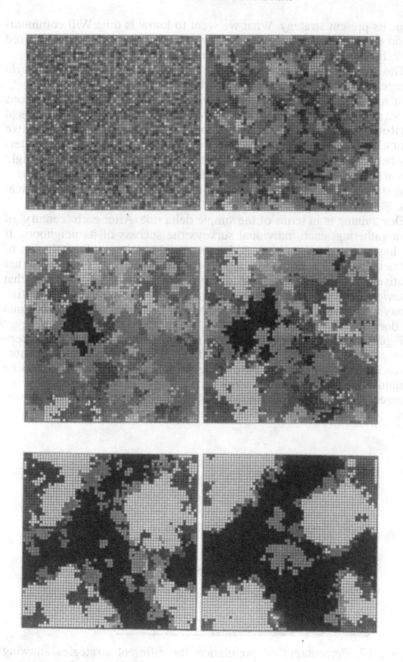

Figure 16. Emergence of communities of perfect communicators in randomized array of perceptrons (three hundred centuries shown).

keeps its present strategy. What we want to know is this: Will communication emerge in such a model? Is it possible, starting with a randomized array, for communities of perfect communicators to arise and grow?

The answer is an emphatic yes. Simple patterns of communication *do* emerge even in an environment this simple.

In a first series of runs, our individuals are the simple perceptrons shown in figure 16: two-layer feed-forward nets, with no hidden nodes at all (Rosenblatt 1959, 1962). In order to keep things simple, moreover, we 'chunk' our weights: each of our twelve weights is chunked at unit intervals between -3.5 and +3.5, giving us a sample space of some sixty-eight billion numerical combinations and a behavioral sample of 38,416 behavioral strategies. Only two of those 38,416 qualify as perfect communicators.

Our training is in terms of the simple delta rule. After each 'century' of point-gathering, each individual surveys the success of its neighbors. If any has garnered more points, it does a partial training on the behavior of that neighbor. For four random inputs, it nudges each of its weights either positively or negatively in the direction required to approximate that behavior. With only four random inputs, of course, training is only partial. It may well be that the result is an individual that differs from the original but does not fully match the 'target'.

Figure 17 shows the evolution of a randomized array of 4,096 perceptrons over the course of three hundred centuries. Perfect communicators are coded in pure black and pure white, with other strategies coded using combinations of background colors and central dots. What one sees is the emergence of two communities of our two forms of perfect communicators.

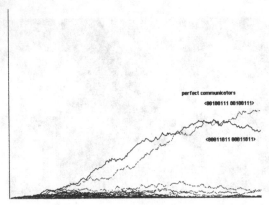

Figure 17. Percentages of population for different strategies showing emergence of perfect communicators in randomized array of perceptrons (three hundred centuries shown).

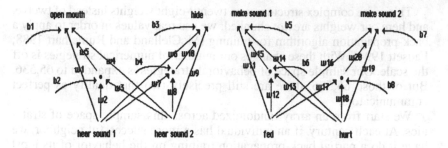

Figure 18. A neural network complete for Boolean connectives in each quadrant.

We can also graph this development in terms of proportions of the population. Figure 18 shows the clear emergence, from an initial random-ized array, of precisely our two perfect communicators.

It has long been known that a neural net of just two layers, like our perceptrons above, is incapable of representing all the Boolean functions: exclusive 'or' and the biconditional are exceptions (Minsky and Papert 1969, 1990). The behavioral range of our perceptrons was thus not quite complete. No individual guided by such a net, for example, could open its mouth just in case it heard precisely one sound and not both. No perceptron could make sound 1 just in case it was either both fed and hurt or neither.

All sixteen Booleans *are* captured in a more complex neural net struc-ture using a limited number of hidden nodes (figure 19). With an eye to computational tractability over an array of 4,096 individuals, we again keep our number of weights as small as possible.

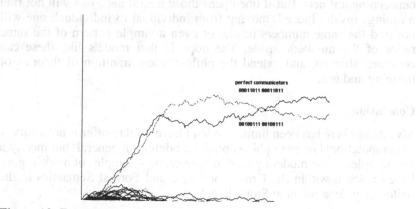

Figure 19. Emergence of perfect communication in a randomized array of back-propagation neural nets. Percentages of population shown for three hundred centuries.

This more complex structure uses twenty-eight weights instead of twelve, and here our weights are *not* chunked: we use real values in order to apply a back-propagation algorithm for training (McClelland and Rumelhart 1988; Fausett 1994). With these changes our number of numerical strategies is off the scale. Our sample space of behavioral strategies is expanded to 65,536. But of those 65,536, there are still precisely two that qualify as perfect communicators.

We start from an array randomized across this sample space of strategies. At each century, if an individual has a more successful neighbor, we have it do a partial back-propagation training on the behavior of its most successful neighbor. This training uses a complete but randomized set of possible inputs for each quadrant and takes the more successful neighbor's behavior as target. Although back-propagation standardly uses a large number of such 'epochs' for training, we deliberately restrict ourselves to a partial training using only a single epoch. Although behavior is nudged in the direction of the most successful neighbor, it will often be the case that the result of a single training matches fully neither the original behavior nor that of the target. The hybridized result, on the other hand, may prove more successful than either on the next round.

Figure 20 shows a typical result with one epoch of training over the course of three hundred centuries.[8]

The core phenomenon of meaning, on models such as these, is simply behavioral coordination between individuals. There is no need for something called a 'meaning' to which utterances or speakers are mysteriously related, and there is no need to portray communication as a transfer of internal representations from one head to another. The fact that sound 1 'means' food and sound 2 'means' predator within a certain community in the model above, for example, is ultimately a fact about coordination between neural nets. But if one opens those neural nets, one will not find meanings inside. Indeed, moving from individual to individual, one will not find the same numbers inside, or even a simple pattern of the same ratios of the numbers inside. The hope is that models like these can continue, sharpen, and extend the philosophical tradition of theories of meaning and use.

Conclusion

My attempt here has been limited. What I have tried to offer is not a survey of computational or even philosophical modeling in general[9] but merely a few samples of the modeling work of one group – samples of modeling we have explored within the Group for Logic and Formal Semantics in the philosophy department at Stony Brook.

---

[8] Although here the two strategies that emerge are our two 'perfect communicators', it is also possible for one to emerge more quickly than the other and to dominate as a result.

[9] For that see Grim 2002.

A more complete account of each batch of work represented here can be found in the sources footnoted. What I hope this quick sampling has communicated is not the details of each exploration but the overall variety of a range of modeling we have been able to pursue in a short period of time. I also hope that some of this conveys the excitement we continue to feel in being able to explore both traditional philosophical questions and new ones with the toolkit of computational models that is now within easy reach.

References

Axelrod, Robert. 1984. *The Evolution of Cooperation*. New York: Basic Books.

———. 1980. "Effective Choice in the Prisoner's Dilemma." *Journal of Conflict Resolution* 24: 3–25.

———, and William Hamilton. 1981. "The Evolution of Cooperation." *Science* 211: 1390–96.

Berlekamp, Ellwyn R., John H. Conway, and Richard K. Guy. 1982. *Winning Ways for Your Mathematical Plays*. Volume 2. London: Academic Press.

Devaney, Robert L. 1989. *An Introduction to Chaotic Dynamical Systems*. Second edition. Redwood City, Calif.: Addison-Wesley.

Fausett, Laurene. 1994. *Fundamentals of Neural Networks*. New York: Prentice Hall.

Grim, Patrick. 1995. "Greater Generosity in the Spatialized Prisoner's Dilemma." *Journal of Theoretical Biology* 173: 353–59.

———. 1996. "Spatialization and Greater Generosity in the Spatialized Prisoner's Dilemma." *BioSystems* 37: 3–17.

———. 2002. "Computational Modeling as a Philosophical Methodology." In *The Blackwell Guide to Philosophy of Information and Computing*, edited by Luciano Floridi. Oxford: Blackwell.

———, Gary Mar, and Paul St. Denis. 1998. *The Philosophical Computer*. Cambridge, Mass.: MIT Press/Bradford Books.

Grim, Patrick, Trina Kokalis, Ali Tafti, and Nicholas Kilb. 2000. "Evolution of Communication in Perfect and Imperfect Worlds." *World Futures: The Journal of General Evolution* 56: 179–97.

Grim, Patrick, Trina Kokalis, and Nicholas Kilb. 2000. "Evolution of Communication with a Spatialized Genetic Algorithm." Research Report 00–01, Group for Logic and Formal Semantics, SUNY at Stony Brook. Forthcoming in *Evolution of Communication*.

Grim, Patrick, Trina Kokalis, Nicholas Kilb, Ali Alai-Tafti, and Paul St. Denis. 2001. "The Emergence of Communication: Some Models for Meaning." Sixteenth Annual Computing and Philosophy Conference, Carnegie Mellon University.

Grim, Patrick, Paul St. Denis, and Trina Kokalis. 2001. "Learning to

Communicate: The Emergence of Signaling in Spatialized Arrays of Neural Nets." Research Report 01–01, Group for Logic and Formal Semantics, SUNY at Stony Brook. Forthcoming in *Adaptive Behavior*.

Hobbes, Thomas. 1997. *Leviathan*. Norton Critical Edition. New York: W. W. Norton. First published in 1651.

Lewis, David. 1969. *Convention*. Harvard University Press.

Lukasiewicz, Jan. 1970. "W obronie Logistyki." In *Z zaganierí logiki I filozofii*. Warsaw: Pauastwowe Wydawnictwo Naukowe, 1961. In *A Wittgenstein Workbook*, translated by Peter Geach. Berkeley: University of California Press.

Mar, Gary, and Patrick Grim. 1991. "Pattern and Chaos: New Images in the Semantics of Paradox." *Nofs* 25: 659–95.

McClelland, J. L., and D. E. Rumelhart. 1994. *Explorations in Parallel Distributed Processing*. Cambridge, Mass.: MIT Press.

Minsky, Martin, and Seymour Papert. 1990. *Perceptrons: An Introduction to Computational Geometry*. Cambridge, Mass.: MIT Press. First published in 1969.

Nowak, Martin, and Karl Sigmund. 1990. "The Evolution of Stochastic Strategies in the Prisoner's Dilemma." *Acta Applicandae Mathematicae* 20: 247–65.

———. 1992. "Tit for Tat in Heterogeneous Populations." *Nature* 355: 250–52.

Poundstone, William. 1985. *The Recursive Universe*. Chicago: Contemporary Books.

Rescher, Nicholas. 1993. *Many-Valued Logic*. New York: McGraw-Hill; Hampshire, U.K.: Gregg Revivals. First published in 1969.

Rosenblatt, Frank. 1959. "Two Theorems of Statistical Separability in the Perceptron." *Mechanization of Thought Processes: Proceedings of a Symposium Held at the National Physical Laboratory, November 1958*. London: HM Stationery Office.

———. 1962. *Principles of Neurodynamics*. New York: Sparton Press.

Sigmund, Karl. 1993. *Games of Life*. Oxford: Oxford University Press.

Skyrms, Brian. 1996. *Evolution of the Social Contract*. Cambridge: Cambridge University Press.

St. Denis, Paul, and Patrick Grim. 1997. "Fractal Images of Formal Systems." *Journal of Philosophical Logic* 26: 181–222.

Stewart, Ian. 1993. "A Partly True Story." *Scientific American* (February): 110–12.

———. 2000. "Logic Quasi-Fractals: A Fractal Guide to Tic-Tac-Toe." *Scientific American* (August): 86–88.

Quine, W. V. O. 1959. *Methods of Logic*. New York: Holt, Rinehart, and Winston.

13

THE *STANFORD ENCYCLOPEDIA OF PHILOSOPHY* A DEVELOPED DYNAMIC REFERENCE WORK

COLIN ALLEN, URI NODELMAN, and EDWARD N. ZALTA

A fundamental problem faced by the general public and the members of an academic discipline in the information age is how to find the most authoritative, comprehensive, and up-to-date information about an important topic. The present information explosion is the source of this problem – more ideas than ever before are being published in print, on CD-ROM, and in a variety of forms on the Internet. One can nowadays use library search engines and web-indexing engines to generate lists of publications and websites about a topic and then access them immediately if they are online. But even limited-area search engines can produce thousands of matches to keywords, and even with new interface tools to narrow the search, one is typically confronted with a list that is not informed by human judgment. If one wants an introduction to a topic that is organized by an expert, or a summary of the current state of research, or a bibliography of print and online works that has been filtered on the basis of informed human judgment, there are few places to turn. One might try a standard reference work, but the main problem with reference works is that they quickly go out of date (even before they are published) and do not reflect the latest advances in research.

So the following questions arise: How can an academic discipline maintain a reference work that introduces the significant topics in the field (for those who wish to learn the basics) but tracks, evaluates, and changes in response to new publications and new research presented in a variety of media (for those with advanced knowledge on a given topic)? How can this be done so that access to the reference work is low-cost, if not free?

Members of our project started thinking about these questions in 1995, and in order to answer them, we developed and implemented the concept of a 'dynamic reference work' (DRW). A DRW is much more than a web-based encyclopedia. The most important features of a DRW are: (1) it provides the authors (who may be scattered in universities all over the world) with electronic access to their entries, so that they can update those entries at any time to reflect advances in research, (2) it provides the subject editors (wherever they are located) with administrative access to those entries and updates, so they can referee them prior to publication (and can add new topics, commission new authors, and so on), and (3) it provides automated tools by which a principal editor can oversee administrative control of (1) and (2) with only a small staff. Thus, on our conception, a DRW includes a highly customized work-flow system by which the members of an entire discipline are empowered collaboratively to write *and maintain* a refereed resource. Such a resource will not only introduce traditional topics in the discipline but also track the (new) ideas that are constantly being published on those topics in a variety of media. With this concept of a DRW, all sorts of new and interesting questions arise concerning how best to design, program, and administer such a resource and work-flow system.

No existing electronic journal or preprint exchange in the sciences or humanities approaches this concept in scope. Electronic journals (1) typically do not update the articles they publish, (2) do not aim to publish articles on a comprehensive set of topics but rather, for the most part, publish articles that are arbitrarily submitted by the members of the profession, (3) typically serve a narrow audience of specialists, and (4) do not have to deal with the *asynchronous* activity of updating, refereeing, and tracking separate deadlines for entries, because they are published on a *synchronized* schedule. Preprint exchanges not only exhibit features (1), (2), and (3) but also do not referee their publications and so need not incorporate a work-flow system that handles the asynchronous referee process that occurs between upload and publication in a DRW. None of this is to say that electronic journals and preprint exchanges have a faulty design, but a DRW is a distinctive new kind of publication that represents a new digital-library concept.

Although commercial publishers have built web-based reference works and claim that they are dynamic, they lack some of the principal design features of a DRW, namely: (1) that authors should have electronic access to copies of their entries and be able to modify them, and (2) that subject editors and the principal editor should have electronic access to the encyclopedia databases and unrefereed entries, so that they can directly carry on the task of adding and commissioning new entries, refereeing entries and updates, and so forth. These commercial publishers typically do not give academics accounts on their computers or access to their databases. Instead, the authors and editors must provide or referee content by first interacting with the staff of the publishing house (managing editors, copy editors, computer web specialists, computer markup specialists, and others) before changes to the encyclopedia can be made public. On our model, the *publishing house* becomes inessential to the process of maintaining a DRW. Academics have direct electronic access to the entries, and they can engage and manage the process of writing, refereeing, and updating entries without intermediaries.

Our implementation of a DRW is embodied by the *Stanford Encyclopedia of Philosophy* (SEP), <http://plato.stanford.edu/>. In the remainder of this article, we document this particular DRW and then discuss some of the outstanding questions and problems it faces.

The Implementation of a Dynamic Reference Work

The SEP first came online in September 1995 with two entries! Since then we have designed a work-flow system that attempts to maximize efficiency among those involved in its production. The most important parts of this system are the password-protected web interfaces to the central server, which can be accessed by any author or subject editor or by the principal editors from anywhere in the world there is a computer with an Internet connection.[1]

The web interface for authors allows them to download our HTML templates, to upload their new entries into a private area of our web server, and to edit remotely copies of their entries stored in this private area. So if an author is lecturing outside her university and encounters a reader of her entry who points out an error or omission, she can sit down at the next net-connected computer (possibly at an Internet cafe), contact the Stanford server using the machine's web browser, and, after supplying her ID and password, remotely edit the content of her piece and submit it for editorial review. The web interface for subject editors allows them to enter new topics, commission authors for those topics, referee and comment on entries and updates submitted for review, and communicate their decisions to the editor. So, for example, if a subject editor is visiting another university and learns by e-mail that an entry has been revised and submitted for review (see the discussion of our tracking and reminder system below), she can use a web browser to log onto the subject editor's web interface, display the original and revised versions of the entry side by side *with the differences highlighted*, easily determine where the changes are located, referee them, and then accept or reject the revised version.

The principal editor also has a special, secure web interface, by which this collaborative process is administered. The principal editor can easily add people to the project, add entries to the database, assign editorial

---

[1] These web interfaces, and the file download and file upload capacities that they enable, are the principal enhancements we have made to the SEP since the publication of "A Solution to the Problem of Updating Encyclopedias," by E. Hammer and E. Zalta, in *Computers and the Humanities* 31, no. 1 (1997): 47–60. When that essay was published, the SEP still used an ftp-based file-upload system. We gave authors system accounts on our Unix server, linked their home directories into webspace, and allowed authors to transfer their files by ftp to our server. Subsequent to the 1997 article, however, when browser-based file upload had become a widely adopted and supported standard, we switched to the new technology. Authors and subject editors no longer needed system accounts on our Unix server, and indeed we determined that maintaining Unix accounts for all participants would introduce problems of scale when dealing with hundreds of accounts. Furthermore, we improved security on our machine by deleting those accounts. Instead, authors were given passwords for the browser-based file uploads. Moreover, subsequent to the 1997 paper, we distinguished a private "upload-space" (which includes "revision space") from our public "web-space." The former contains private copies of the entries accessible only to authenti-cated users, so that newly uploaded entries, and newly revised entries, do not become publicly viewable until after they have passed through the referee process.

control for entries to the subject editors, issue invitations, track deadlines (for new entries and for updates), and publish entries and updates when they are ready. Many of these things can be done with just the press of a few electronic buttons. For example, when a subject editor submits (through her web interface) a suggestion to commission an author for a particular topic, the suggestion gets entered onto a database, and the principal editor is notified and prompted to log onto his web interface. He simply hits the New Invitation button, selects the entry in question, and is then prompted to invite the person listed in the database for that entry by hitting the Invite button.

Finally, we should mention that we have designed and implemented a web interface for *prospective* authors. When a prospective author receives an invitation, he or she is directed to log onto a special web interface to obtain information about the project, to set up an account with us if he or she plans to accept, and to set a deadline of up to a year for completing the entry (or else write to us with a counterproposal).

These 'front-end' web interfaces supply data to the 'back-end' processing programs and databases in our system. In particular, actions taken, and information entered, by authors, editors, and prospects are communicated to our tracking and logging system. This system can identify the state of any given entry, recognize who now owes work on an entry and which deadlines have or have not been met, and pass this information to our automated e-mail reminder system, which has recently been developed, initialized, and put into continuous operation. When an entry changes state and another person must now act to continue the publication process, the reminder system will prompt this person about what needs to be done and by when. It will continue to send reminders (on a fixed, inoffensive schedule) until the work is done (or notify the principal editor that all reminders have been ignored and that human intervention needs to take place). Finally, when any entry or substantive revision is published, the entry is scheduled for revision within three to five years (depending on how swiftly the field moves). Actually, some authors update entries once a year, but all authors are notified by our reminder system well in advance of any scheduled revision.

The use of these web and computer technologies offers considerable savings over more traditional publishing methods, and it has enabled us to develop the *Encyclopedia* with a small staff and budget. The importance of this project for our profession cannot be overstated. As new ideas in logic, ethics, political philosophy, philosophy of science, philosophy of cognitive science, and so forth, are published in books and journals of philosophy, both in print and on the web, the SEP provides a rational and efficient system by which the new information is assimilated, digested, and disseminated in entries that are *responsive* to new research.

Here is a basic quantitative analysis of the effectiveness of our design, which is justified by what we have said above. Consider first the fact that there was a thirty-year gap between philosophy encyclopedias (the *Macmillan Encyclopedia* was published in 1967, the *Routledge Encyclopedia* in 1998). So there was no up-to-date encyclopedia for at least twenty-five years (9,125 days). By contrast, a typical *Stanford Encyclopedia* author is regularly visiting the library to read journals or receiving new journal issues at her office. As soon as she realizes that a recently published article advances the topic of her *Encyclopedia* entry, she can use her computer to call up the *Encyclopedia* server and modify her entry accordingly (maybe by adding a paragraph and a bibliographic item). In principle, this could occur on the day that an article is published, and in some cases she might even have advance knowledge of the publication if the author has sent her a preprint. The next day, assuming that the change is a substantive one, the relevant subject editor(s) will be notified that the revision must be refereed. Suppose it takes a week to referee the minor changes to her entry. Then we have reduced the length of time required for the update process from about nine thousand days to about nine days, or by three orders of magnitude. Even if it were to take up to a year for a new idea (in a book, say) to become reflected in the *Encyclopedia* after the new idea is published, that would still constitute a twenty-five-fold decrease in the length of time it takes for a philosophy reference work to reflect the advance.

We should mention two other features of the SEP that should be part of any DRW. The first is the fact that authors are encouraged to write nested, as opposed to linear, documents. That is, we encourage our authors to put highly technical, scholarly, or detailed information into supplementary documents and to link these to the main part of the entry. These supplementary documents can have supplements as well, and so forth, and the reader can then choose the level of detail he or she wishes to explore. Such nested entries become useful to a wider range of readers – intelligent undergraduates should be able to get through the main entry by skipping the links to the supplements, while graduate students and colleagues may skip the basics and follow the links to the supplementary documents, to find the cutting-edge material.

The second noteworthy feature concerns archiving. For purposes of citation, a DRW is a moving target, because the entries are always being corrected, updated, improved, and so on. It is difficult to cite such a moving target. For example, a reader might quote a passage from a DRW entry in a research article, and after publishing the research article, discover that the author of the DRW entry has altered the passage in question. To solve this problem, we make quarterly archives of the SEP. On the equinoxes and solstices, we make an electronic copy or 'snapshot' of the entire *Encyclopedia* as it exists on that day and link that complete copy to our special Archives page. We explain to users that the proper way to cite an SEP entry is to cite the most recent archived version. These archived versions will not be updated or changed in any way, and so scholars can rest assured that the passages they quote will be available for scholarly purposes. Note that every entry in the SEP contains a section called Other Internet Resources, which contains links to offsite web-based material, and these links may eventually break in the archived entries (especially if the links do not point to similarly archived material). That is a danger of the web. But we do attempt to minimize the problem, however. We have designed and programmed a 'link-rot' detection system that automatically notifies the authors whenever links break in the dynamic versions of their SEP entries. The authors are asked to revise or delete the link.

Statistics about the SEP as a Dynamic Reference Work

As of September 21, 2001, the SEP had 213 entries online. We had sixty-nine subject editors overseeing 513 authors currently working on a total of six hundred commissioned entries. More than 10 percent of our entries make use of some hierarchical document structure (that is, they involve more documents than simply a main text and a footnotes page), and just under half of our entries have been updated since they were first published.

The rate at which we commission entries increased by a factor of three, from about five per month in 1999 and 2000 to about fifteen per month in 2001. (See figure 1.) During the same period, our publishing rate increased by a factor of six, from about 1.5 entries per month in 1999 to nine entries per month in 2001. (See figure 2.) The average length of our entries also increased, from approximately 6,800 words per entry in September 1998 to 8,900 words per entry currently.[2] We estimate that, in print, the current version of the SEP would fill more than three thousand pages.[3]

Between September 1997 and September 2001, the content of the SEP grew from about three megabytes to about twenty-six megabytes. (See figure 3.) During that same period our average accesses per month increased by an order of magnitude, from about five thousand to fifty-seven thousand. (See figure 4.)

Why the SEP (and other DRWs) Should Be Free

After six years of operation and numerous discussions with authors and subject editors involved in our project, with faculty members and colleagues around the world, with publishers, and with university librari-

---

[2] These word counts are slightly inflated due to the presence of HTML tags in the text. We estimate that the tags add about three hundred to five hundred words per entry.

[3] This estimate is based on an assumption of six hundred words per page.

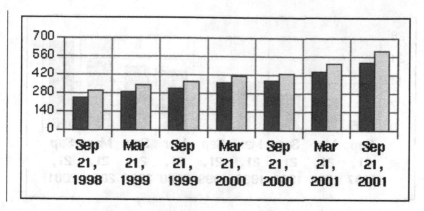

Figure 1. Number of authors/commissioned entries

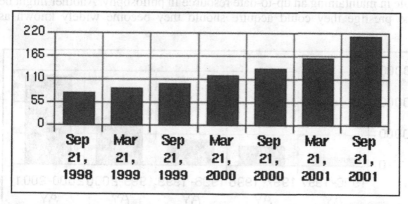

Figure 2. Number of entries online

ans, we have come to the conclusion that if the means are possible, the SEP should try to remain a free resource. One may think of this as an idealistic goal, but there are several problems associated with subscription-based, access-restricted models for funding the SEP. We discuss some of these problems in what follows.

The first problem arises from the fact that academics who write entries for scholarly reference works traditionally receive a fee for their efforts. They are, after all, providing a service to the publisher. Authors of SEP entries, however, are volunteering their time. There are various reasons why they do this. One might be the fact that they will reach a large number of readers. (So long as the SEP remains free or low-cost, it will have a large readership.) Another might be the intellectual obligation academics might feel to contribute to the profession and world at large by playing a

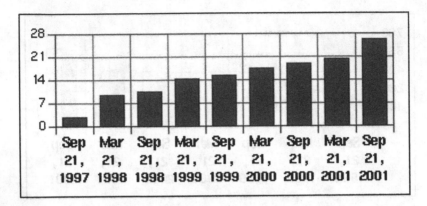

Figure 3. Content in megabytes

role in maintaining an up-to-date resource in philosophy. Another might be the prestige they could acquire should they become widely known as

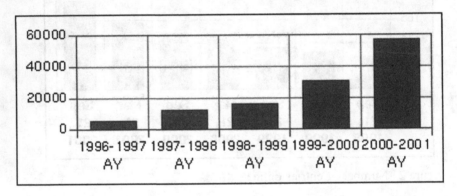

Figure 4. Weekly average number of users (excluding mirror sites)

experts on a certain topic or by becoming associated with a Stanford University project. And, of course, authors may be motivated by the fact that they can list these entries as invited publications on their curricula vitae, which play an important role in promotion, hiring, and tenure.

This volunteer arrangement, however, might become compromised if the SEP were required to charge subscriptions in the attempt to make a profit. Even if the SEP charged subscriptions at a rate that simply recovers costs, authors might suggest both that we should recognize their efforts as part of that cost and that we should therefore increase prices a little so as to collect enough revenue to distribute royalties to the authors. The situation is complicated, moreover, by the fact that authors might *in addition* argue that maintaining an entry is also a service. In any year they update their entry, they might be owed remuneration. As we scale up to more than a thousand authors, this problem becomes even more acute. And a final wrinkle is the fact that our authors currently provide us with entries formatted in HTML and nearly 'web-ready'. If we charged a subscription, if only to recover costs, authors might claim that we are *offloading our typesetting costs onto them*. It would be best to sidestep all of these questions by finding a model in which the SEP could remain a free resource and our authors never feel that they are being exploited.

A second, somewhat similar, problem concerns our subject editors. They are constantly suggesting new entries, commissioning entries, and refereeing entries and updates. If we required subscriptions for access to the SEP, if only to recover costs, the subject editors might argue that their services should be recognized as part of those costs, and that we should increase our subscription rates to pay them accordingly. This problem is magnified by the fact that the work involved in maintaining a dynamic reference work continues from year to year, with no fixed end point, and it is complicated by the fact that some subject editors have higher work loads than others.

A third problem concerns the difficulty of finding a subscription model that everyone can live with. University libraries and other institutions are reluctant to pay for something the rest of the world can get for free. In consequence, it would be difficult to implement a model in which these institutions would pay a subscription fee while everyone else would be allowed free access. So a subscription model might eventually force us to require everyone to *subscribe* (albeit at appropriately proportional rates).

A fourth problem is the fact that many deserving groups of people would be disenfranchised by a subscription model. Small colleges, public libraries, and K-12 school libraries usually cannot afford even a modest subscription fee. Moreover, people who access the web from home over an ISP seem reluctant to pay for content, much less for a subscription to a *philosophy* encyclopedia. And last, but not least, our users in the developing parts of the world would become disenfranchised; this includes university students and colleagues as well as lay persons in developing countries. As the SEP has been accessed by users in more than 150 countries around the world, academics associated with the project as authors would lose a significant base of readers.

A fifth problem concerns search engines. When people search Google, Yahoo, and the like for philosophical topics, our pages figure prominently in the results. That is because our pages are free to everyone, and web-indexing spiders can access and index our pages. But on a subscription model, our server would restrict access to those who pay. This would make it nearly impossible for search engines to index our site, and so people would not find the SEP pages when they conduct web searches. Even if some arrangement could be made to allow indexing spiders to index our site, access restrictions would make the links returned by a search engine useless to the majority of web users and make it difficult, and pointless, for nonpaying users and institutions to create links to our site on their own pages. But without those thousands of links to the entries on our site, the prominence of our pages in the results of certain web search engines (such as Google) would greatly diminish, because those search engines often prioritize the web sites that are returned as matches to keyword searches by the number of cross-referencing links to those web sites that exist on the web.

A sixth problem concerns our mirror sites. Currently, the SEP has mirror sites (University of Sydney, University of Amsterdam, and University of Leeds), which perform several functions. (1) They guarantee our users access to *Encyclopedia* pages when the Stanford server temporarily goes down. If anything happens to our server, our readers use a mirror site until we fix the problem and reboot. (2) These mirror sites give our users in other parts of the world faster access to our pages. (3) The mirror sites provide very important layers of digital preservation for the SEP pages by keeping complete copies of our data. These institutions completely under-write the costs of the mirror sites and provide this service for free. But this arrangement would come to an end if we started charging a subscription for our site. On a subscription model, we would either have to pay the mirror sites for their costs in manpower, equipment, and overhead (assum-ing that some monetary arrangement could be made), or have to face the much more expensive proposition of running 'mission critical' servers that provide fast, worldwide access twenty-four hours of every day. (If institu-tions had to pay for the SEP, they would expect a level of service at least comparable to that which they now enjoy.)

A seventh problem is the fact that if the SEP were forced to rely on cost recovery to survive, the decisions that are now made *in part* on the basis of the interests of the profession may have to be made instead on the basis of cost recovery *alone*. This might have a negative impact on the quality of the *Encyclopedia*, such as placing strict word limits on the length of entries and bibliographies (to save disk space or to stay within bandwidth limitations), banner advertising, links to online booksellers, and so on. Links to booksellers, for example, would compromise the integrity of the *Encyclopedia*, as a user might wonder whether the link to an online book-seller is present because of the merits of the book or because the online bookseller is kicking back some of the profits to the publisher.

In addition to these seven problems that would arise if the SEP adopted an access-restricted, subscription-based, cost-recovery model, we believe that there is a positive reason for remaining free – namely, that it would be an outstanding legacy for the SEP, and the profession as a whole, if the SEP could provide both academics and nonacademics around the world with a free resource through which they could satisfy their intellectual curiosity from an authoritative source on philosophical questions of all kinds, in particular those concerning the human condition.

Costs of the SEP

It does of course cost money to produce the SEP, and if it is to remain free, those costs will somehow have to be underwritten. Currently the SEP costs are underwritten by a grant from the NSF, which will fund the SEP until September 2003.[4]

The NSF grant pays for a principal editor (working 50 percent time), a consultant Perl programmer (working 33 percent time), an associate programmer (working 20 percent time), an assistant programmer (working 10 percent time), and an administrative assistant (working 10 percent time).[5] Clearly, this adds up to a tiny staff, equal to one person at 100 percent employment and one at 23 percent employment. We believe our accomplishments are significant in light of our tiny staff, though it should be mentioned that these accomplishments are made possible by the fact that our personnel regularly put in longer hours than official records indicate. We are currently seeking grant money to hire a business consultant to determine exactly what costs are required to run and staff the SEP and what funding models are available to ensure the SEP's long-term survival.

We believe that the SEP could evolve into an even greater publication (with entries of even higher quality, with fewer typographical and other errors, and so on) if it had more adequate staffing. It does not strike us as unreasonable to think that the SEP should, for the long term, require a principal editor at 50 percent time, an administrative assistant at 50 percent time, a programmer at 50 percent time, and an HTML (XML) copy editor at 50 percent time. With an administrative assistant and HTML copy editor working at these levels, the principal editor would be relieved of certain routine tasks and could concentrate more on matters of content, on supporting subject editors for those subjects of the SEP which still are not very far along, and so forth. Similarly, with a programmer working half time, we could reengineer many of the compromises we made in designing and programming our work-flow system, we could adapt the SEP to new technologies that improve web-based publication, and so on. Although our estimate of these long-term staffing requirements for the

---

[4] The NSF grant (#IIS-9981549) was made possible by a significant financial contribution from the NEH. The NEH previously funded the *Encyclopedia* project from September 1998 to August 2000 (#PA-23167-98). From September 1995 to August 1998, the SEP was funded by seed money from the Center for the Study of Language and Information (Stanford University), where the project was conceived and developed. CSLI still contributes some cost-sharing funds under the terms of the NSF grant.

[5] After we complete the task of programming the essential work-flow systems underlying the SEP, there will be less programming time required. If we are to keep up with changes in technology, however, programming time will always be necessary.

SEP will need to be subjected to a strict business analysis, it should give one some idea of the money that the SEP will need for salaries. Once money for overhead costs (including hardware and software) are included in the equation, it should be clear that the SEP will have to find sources of income if it hopes to remain a free publication after the NSF grant expires.

Future Challenges

Clearly, then, the principal future challenge for the SEP is to find a source of income that will enable it to continue to be universally accessible. In this concluding section, we discuss this challenge, along with two others.

Funding Models

We shall, of course, attempt to raise an endowment. Indeed, we plan to submit grant proposals to foundations, asking them to give us money to hire a fund-raiser for a year, at full or half time. But we anticipate that it could take up to ten years to raise an endowment large enough to cover our operating costs.[6]

Our present focus, however, is to explore funding models intermediate between universal free access and full-fledged subscription models. For example, we plan to study the following two models, both of which might be adopted without incurring all of the problems we outlined above concerning access-restricted models.

Voluntary Archive-Acquisition Program. On this income-producing plan, the SEP would remain free to everyone, but institutions subscribing to this program would be entitled to download and store our four quarterly archives each year they subscribe. (Those who choose not to pay would still be able to access all our pages.) There would be two advantages to subscribing: (a) the institutions that subscribed would be entitled to serve their copies of the archives whenever our server and its mirror sites are down, and (b) should the SEP project ever cease to exist, these institutions would own, and be able to circulate locally, copies of our archives even though our servers are no longer active. For an extra fee, we could enhance this service by burning and distributing yearly CDs of the archives for that year.

Archive-Access Program. On this income-producing plan, *only* those institutions and users who pay a yearly subscription fee would be able to access our archives for that year. The SEP's dynamic entries (which are always changing) would still be available free to everyone. However, because

---

[6] Remember that if endowments are managed properly and make 8 to10 percent a year, then approximately half of the 8 to 10 percent would be returned to the SEP and the remaining half of the 8 to 10 percent would be reinvested.

only fixed, archived versions of the entries can be cited, the motivation for joining this program would be clear to scholars and librarians. On this plan, we could again create the following levels of service: (a) for a basic yearly fee, users at subscribing institutions would receive web-based access to the archives on our server, (b) for a higher fee, the institution would obtain the right to download and store the archives, and (c) for an even higher, but still moderate, fee, we would burn and distribute CDs to the subscribers.

We plan to investigate these and other models by which income could be raised. They strike us as offering the best hope of raising money in a way that addresses the problems we discussed earlier.

Institutional Home
The question of funding models is connected to the question of where the best institutional home for the *Encyclopedia* is. We believe that the SEP is better off in an academic environment than in the hands of a commercial or even nonprofit publisher or publishing house. Normally there would be three good reasons for joining forces with a nonprofit, university-based publisher – namely, that such a publisher would: (1) offer expertise in the business of producing publishable material, (2) offer a more stable institutional home, and (3) have the mechanism for marketing and collecting revenues for the materials it produces. We consider these in turn.

Expertise. We are not aware of any nonprofit publishers with the kind of expertise that already resides with the SEP project. Since 1995, we have been perfecting our model for dynamic reference works that can be run on a low-cost basis. Grants from the NEH and NSF for 1998 to 2003 have given us the financial resources to program a work-flow system that would compare, on the open market, to off-the-shelf 'web content management' software systems costing \$200,000 to \$300,000. The *Stanford Encyclopedia of Philosophy* has been successful to date as a project designed, programmed, and run by academics (not specialists) who have acquired the tools necessary to make use of the power that computers, the Internet, and web-based technologies provide. In particular, the most highly technical positions required by our project, Unix system administration and Perl/CGI programming, are filled by *academics* working part time.

A More Stable Institutional Home. Currently the SEP is published at Stanford University's Center for the Study of Language and Information (CSLI), and the Stanford Department of Philosophy serves as its Advisory Board. Our analysis suggests that the best institutional home for the SEP is an academic setting of this kind. Numerous reasons for this are readily apparent. An institution like CSLI or an academic philosophy department would have a more intimate and direct concern for the academic excellence of the project and for safeguarding it as a resource for the profession. Such a concern might not be shared by nonprofit, university-based publishers, since traditionally they let titles go out of print. Moreover, a philosophy department or research institute has a concern for educating its graduate students, and these students could play various and vital roles in the SEP project.

Many graduate students in philosophy have a background in mathematics and computer science. They arrive at graduate school with enough knowledge about Unix, web servers, and so on, to work on or consult for our project. There are two main ways in which these students can provide a steady stream of innovative ideas for the future of the project. First, as paid, part-time members of the SEP staff. Graduate students would make excellent part-time staff. As such, they could (a) help subject editors plan and commission entries (thereby becoming known to distinguished members of the profession outside their home institution), (b) acquire and use HTML or XML skills and help the *Encyclopedia* with content markup, (c) acquire and use their knowledge of Unix and web servers to help administer the project (thereby preparing them to use those skills in their later academic life), and (d) work as office staff, handling correspondence with the authors and relieving the principal editor of routine tasks in the administration of the *Encyclopedia*.

Second, work on the SEP could be made part of the graduate curriculum. A philosophy department, without exploiting the graduate students in any way, could create a one-hour weekly proseminar for all first-year graduate students. Each week, the students would be required to read and report on one article from the SEP in their field or in a related field. In this way, graduate students would enhance their breadth and analytic skills as philosophers, improve their writing skills by focusing on whether entries are well written from a pedagogical point of view, and suggest ways to improve and/or update entries. They could use their talents for web-based searching to identify whether any related material on the web should be linked into the entries they consider.

Finally, it is important to note that a research institute like CSLI can be an important collaborator in this project. For instance, CSLI has researchers with expertise in linguistics, computation, data-mining, and so forth. It also has a highly skilled technical staff, which can deal with any technical issues (such as those connected with server operation, backup, and so on) that might go beyond the expertise of a typical academic department.

Marketing and Income Collection. Of course, if we can find a way to keep the SEP a free resource, then the fact that a publisher offers marketing and income collection becomes moot. As we have seen, however, we may be forced to adopt an operating model intermediate between that of keeping the *Encyclopedia* free and requiring universal subscription. Such models would require some marketing and income collection.

Basic marketing would not be problematic for Stanford's Department of Philosophy and CSLI. When our NSF grant runs out in 2003, the SEP will have been on the web as a *free* resource for eight years. During our first six years, we have become well-known throughout the philosophical world (both academic and nonacademic), and many thousands of individuals know about, read, and rely on our pages. If our service were to be shut off or restricted in some way, libraries and other institutions would certainly hear about it from their constituents. Moreover, most of our five-hundred-odd authors and seventy subject editors would notify the libraries at their own institutions. In addition, the fact that the SEP is a free resource means that there are thousands (if not tens of thousands) of links to our web pages. If we were to start charging subscriptions for access, these links would all end up in an *advertisement to subscribe*, as soon as anyone followed a link.

Income collection could also be easily accomplished, and could be done by using a university-based eCommerce center, such as the one at Stanford University.[7] These eCommerce centers can provide subscription services via the web and can handle subscription payments for departments that create journals or other publications for sale. The eCommerce centers are relatively inexpensive to use.

Finally, we should mention that if a more sophisticated system for marketing and income collection is required, an alliance with the *nonprofit* Philosophy Documentation Center might be possible. Discussions to this effect have already taken place.

A Move to Newer Technologies?
The SEP was designed to run on proven, free technologies. Of course, since we began the project in 1995, some of these technologies are now legacies. But in many cases we chose to use certain technologies over others because they made the most sense given our budgetary constraints. We did have to make compromises in some cases.

We use HTML rather than XML for entry markup. We do not use a heavy-duty 'application server technology', as performance is excellent with our Apache server and Perl/CGI scripts. We do not rely on a heavy-duty data-base technology (such as Oracle), because our main data base has (or will have) only one thousand to two thousand records, as opposed to hundreds of thousands of records. We do not rely on Java or Javascript, although our authors are free to use it in their entries if needed. We avoid frames on the main pages of the *Encyclopedia*, although we use them in the web interfaces.

Those who follow trends on the web, however, will know that this implementation is relatively low tech. But this low-tech approach does have several advantages. One is that the system can run on any PC running

---

[7] See <http://www.stanford.edu/group/itss-ccs/project/ecommerce/>.

a free Unix-based operating system, such as Linux or FreeBSD. Our low-tech approach also does not require that we purchase any licenses or that we become dependent on any commercial software. Just as important is the advantage that philosophers with limited computer savvy and expertise, as well as philosophers in parts of the world where access to high-tech or up-to-date computer systems may be limited, can participate in the collaborative production of our DRW.

In addition, the newer technologies always seem to increase the costs of production (they often require specialized personnel, for example), and until we are satisfied that the benefits outweigh the costs, we shall exercise caution when considering the latest technologies. But eventually some of these newer technologies may supersede the older ones and the SEP will have to make the needed adjustments, assuming it is in a financial position to do so.

The question we are asked most frequently is, When do you plan to adopt XML as a markup standard? XML is now highly touted as the markup language to use in web publications. As a markup language, XML offers some serious advantages over HTML. For one thing, the tags are constructed on the basis of the kinds of content that appear in a document. For example, in HTML one would format book titles using italicizing tags, such as <i> . . . </i> or <em> . . . </em>. But in XML one could format book titles with the tag <booktitle> . . . </booktitle>. This would allow one to search the SEP in more sophisticated ways. One could tell the search engine to search only keywords in the <booktitle> environment, whereas in HTML there is no way for a search engine to distinguish a keyword found in a book title from one found in other italicized environments, such as emphasized text or foreign words and phrases.

Another advantage of XML is the promise of more sophisticated mathematical and logical formatting. HTML has only weak resources for formatting sophisticated mathematical and logical notation. There is some promise that MathML (a formatting language that is a special instance of XML), when supported by MathML-aware browsers, will give web publishers the ability to publish professional-looking mathematical and logical notation.

These virtues of XML come at a cost, however. First and foremost is the fact that because XML is a more sophisticated markup language, the costs of production rise significantly when taking proper advantage of XML's extended capabilities. Currently our authors provide us with nearly 'web-ready' HTML documents, which they produce with freely available HTML-editing software.[8] Indeed, they are free to use any HTML editor to compose their entries – we do not want to force all authors to have to learn

---

[8] It must be mentioned, however, that we always have to spend time bringing their documents into compliance with international standards (e.g., removing proprietary HTML formatting codes that the HTML-creation software introduces) and making them consistent with our other entries.

and/or use the same composition software. Until XML-editing software tools become widely available and easily configurable, our authors will not be willing to spend extra time using all the new tags provided by XML to format their entries. (For example, authors will understandably be reluctant to familiarize themselves with all the special tags, such as <booktitle>, and use them in their entries.)

There may be some ways, however, to ameliorate these costs. Suppose, for example, that there was a free software application that the authors could use to help them to format entries in XML graphically and automatically without learning the new tags. For example, whenever an author wanted to insert a new item into the bibliography, such a piece of software would, in response to a click on an 'Add Bibliography Citation' menu item, pop up a window containing all the relevant fields (book title, article title, author name, date, city, publisher, and so on) of a typical citation. When an author inserted the information into these fields, the software would then mark the information with the appropriate XML tags in the source file. There is very little extra cost to the authors, as they have to type in the information in the bibliography citations anyway. Of course, such a piece of software would be expensive to design, produce, and support on the major computer platforms. Until such software is widely and freely available or the SEP has the financial resources to hire XML-markup specialists, the move to XML will be problematic.

There is also a second problem with XML, which has to do with the fact that our authors now have electronic access to their entries and can keep them up to date. As we mentioned earlier, authors can use their browser to contact the SEP's server through a web interface. When they activate our 'Make Changes' function, their browser will divide up a private copy of the entry into sections and, for each section, display both the rendered HTML and an editing box to the HTML source file. An author can then edit the HTML source file and redraw the screen to see that the HTML is rendered correctly. It is not difficult for authors to read past the HTML formatting tags to edit the text they wish to change, or even to add basic HTML formatting to their updated text.

But this procedure becomes more difficult in XML and especially MathML. XML source files are much more highly formatted than HTML source files. A simple equation like $2^2 + x = 8$ requires numerous MathML formatting tags, and it is much harder to read past these tags to find and edit the text. (MathML was designed with the idea that authors would never actually edit the source file but always edit the file through a graphical interface.) So a move to XML would make it more difficult for our authors to update their entries, as the source files would become much more difficult to edit. Again, there may be a way to get around this through a Java-based applet/servlet system, which presents authors, no matter where they are located in the world, with a graphical editing interface to their XML source files. But such a Java applet/servlet combination is extremely difficult to program so that it works with all combinations of computer architectures, operating systems (and their different versions), web browsers (and their different versions), and so forth. The costs would be exorbitant for a project that hopes to keep costs to a minimum. It is doubtful that a single full-time programmer could design, produce, maintain, and support such an application. In consequence, a premature move to XML would interfere with the ease of scholarly communication that the SEP now enjoys.

As one can see, then, there are many challenges facing the SEP. We have a system that works reasonably well now, and we are working to secure the SEP's future for the long term.

14

CULTURES IN COLLISION:
PHILOSOPHICAL LESSONS FROM COMPUTER-MEDIATED
COMMUNICATION

CHARLES ESS

Philosophers have documented numerous ways in which the rise of computing technologies has provided us with new kinds of laboratories in which we can empirically test various theories and hypotheses. So Patrick Grim and his colleagues have programmed individual cellular automata to pursue different moral strategies in the framework of the classic "Prisoner's Dilemma." Their computations demonstrate that a strategy of "forgiving tit-for-tat" is the most adaptive from an evolutionary standpoint – that is, this strategy, over a run of hundreds of generations, leads to the dominance of cells so programmed.[1] The computer thus provides a controlled experimental environment in which philosophers can model and critically evaluate specific ethical and political claims.

Similarly, computer-mediated communication (CMC) has created for philosophers new sorts of laboratories in which *communicative* behaviors can be observed, documented, and analyzed. For example, Susan Herring has demonstrated that gender remains largely apparent in e-mail, listservs, chatrooms, USENET postings, and so on, contrary to the prevailing belief that CMC provides users with a "gender-blind" environment that might liberate women from the patriarchal patterns of discrimination and oppression in the "face-to-face" world. Indeed, her analyses have suggested that anonymity and pseudo-anonymity in CMC environments work as a social amplifier – such that aggressive patterns of communication characteristic of males (including flaming and sexual harassment) become even more predominant (see especially Herring 1996, 1999; Hall 1996). Herring's work corrects on *empirical* grounds initially *philosophical* claims concerning the social and political consequences of CMC technologies – claims that otherwise rested largely on theoretical arguments with comparatively little empirical support.

More broadly, computer networks have made interaction between

---

[1] Grim et al. 1999; Grim et al. 2001: see also the online version of Grim et al. 1999, especially "I.3 Cellular Automata and the 'Evolution of Cooperation': Models in Social and Political Philosophy," <http://www.sunysb.edu/philosophy/faculty/pgrim/intro.html#sec3>.

peoples of different cultures possible on a scale (at the time of this writing, somewhere between 5 and 7 percent of the world's population), scope (the more or less global reach of the Net),[2] and speed never before available. CMC technologies have brought about extraordinary new possibilities for engagement between *cultures* and *peoples*. These same computing technologies further allow philosophers and researchers from such disciplines as sociology, anthropology, communication, political science, human-computer interaction, and so forth, carefully to record, observe, and analyse these engagements. CMC technologies provide us with remarkable new kinds of laboratories for examining philosophical claims concerning culture and its intersections with technology and communicative preferences and styles. Just as the collisions between particles provide physicists with the data they need to determine with greater precision the characteristics of the still smaller particles and forces that underlie the matter they can record in their laboratories, so the cultural collisions apparent and recordable in CMC environments can help us become more explicit and critical about the underlying assumptions of culturally diverse worldviews and communication preferences that otherwise remain tacit and unquestioned.

In other terms, this attention to the philosophical lessons of implementing CMC in diverse cultural contexts is in part the task of philosophy of communication – the conceptual space in which the methods and insights of philosophy and communication theory mutually illuminate one another. Arguably, the twentieth century represents a strong turn toward philosophy of communication, as it includes language analysis and the work of Wittgenstein, speech-act theory (Austin and Searle), and most especially the communicative models of rationality that emerge from Gadamer, Ricoeur, Habermas, Apel, Bahktin, Davidson, and Rorty. Much more modestly, this essay relies on empirical analyses of what happens when CMC technologies are deployed in diverse cultural settings to evaluate critically claims and arguments in philosophy of technology, epistemology, identity, metaphysics, and ethics, and to draw constructive conclusions for philosophy as a practically oriented, interdisciplinary, and global enterprise.

I first examine how a range of cultural encounters occasioned by the

---

[2] The meaning of "global" in this context needs careful qualification. Simply put, the exponential growth of the Internet and the World Wide Web has *not* – so far – led, as enthusiasts like Nicholas Negroponte (1995) hoped, to greater equality in terms of distribution of and access to the expensive infrastructures and technologies needed to support computer-mediated communication. On the contrary, the digital divide – both *within* well-to-do countries like the U.S. (see U.S. Department of Commerce et al. 2000) and *between* the developed vs. the developing nations – appears to be holding steady, if not gradually growing worse. For that, the G8 summit in July 2001 endorsed an "action plan to bridge the digital divide" (see <http://www.dawn.com/2001/07/22/ebr13.htm>) and the World Bank has embarked on an ambitious series of development programs to overcome the digital divide (see, for example, the issue brief "Knowledge for All," <http://www.worldbank.org/html/extdr/pb/pbknowledge.htm>).

implementation of Western CMC technologies in diverse "target" cultures shed light on a central debate in philosophy of technology. I use observations and data on implementing CMC in Japan, the Middle East, South Asia, and among indigenous peoples in the Philippines, South Africa, and Malaysia, to examine critically the claims of *technological instrumentalism* (the claim that technology is "merely a tool," an instrument that carries no culturally dependent values) and *technological determinism* (the claim that once technologies are unleashed, they become autonomous powers whose progress and impacts on human beings and their cultures cannot be steered or stopped). This inquiry makes clear that neither claim is fully supported by the results of implementing CMC systems design in the West in diverse cultures – thus pointing to a middle-ground position of *soft determinism*.

This first thread of inquiry, moreover, provides the data and interdisciplinary theoretical approaches (as taken from communication theory especially) to illuminate philosophical debates concerning the nature of the self and epistemology. In each of these areas, hypotheses and assumptions can be tested in light of the distinctive new forms of praxis and empirical data made available by CMC environments. Such praxis and data serve for philosophical claims as a strong analogue of the empirical researchers' experimental data – that is, empirical evidence from repeatable experiences that can at least disconfirm initial hypotheses (as in the example of Susan Herring's work), if not support one claim over another (as Grim et al. have done).

These examples of philosophical inquiry as practiced in the context of CMC as a kind of philosophical laboratory illustrate significant ways in which CMC constitutes both a crucial component of philosophy of communication and another strand of the "computational turn" in philosophy thematic of this collection of essays. In addition, these examples suggest two crucial lessons for philosophers. First, contrary to the pressures toward specialization within philosophy narrowly construed, it should be clear that the sorts of insights reviewed here are simply not possible without engaging in forthrightly interdisciplinary collaborations. The computational turn hence requires philosophers to reach beyond the boundaries of their own specialties and discipline – as is already exemplified in philosophy of communication as an interdisciplinary field – in order better to understand, through collaboration with communication theorists, social scientists, cultural-studies scholars, and others, just what philosophical insights may be gleaned from these new technologies. Second, the interdisciplinary engagement of philosophers in the emerging design and implementation of CMC technologies is critical if we are to contribute to a global ethic and a Socratic form of education that are increasingly required if users of these new technologies are to avoid a form of computer-mediated colonization – the result of our ethnocentrically assuming that all peoples must think, believe, and communicate as we do via CMC. Such an ethic and education are essential conditions for the Stoic conception of the *cosmopolitans*, the "citizens of the world" who deeply understand and can maneuver comfortably among multiple cultural worldviews and communicative preferences. These cosmopolitans engage with one another via global forms of CMC in ways that preserve and enhance foundations of culture, rather than simply colonize them into a single homogeneity. A philosophically shaped global ethic and a Socratic education are necessary, in short, for a genuinely intercultural electronic global village.

Philosophy of Technology: Technological Determinism, Technological Instrumentalism

The iconic image of an electronic global village rests on two specific assumptions – ones often debated in philosophy of technology. On the one hand, McLuhan (1965) presumed that new media technologies were themselves morally neutral or value free. Such technologies are presumed neither to embed nor to foster any given set of ethical or cultural values: as morally neutral, the only ethical questions that may arise concern the ends, or goals, to which they serve as means. This view of technology is discussed by philosophers of technology as *technological instrumentalism*.[3] On the other hand, a central justification for the use of CMC technologies in an emerging global village is the claim that these technologies facilitate a *democratizing* form of communication – that is, communication that flattens traditional hierarchies, expands freedom of individual expression, and gives everyone a voice in a global society.[4] This justification seems to presuppose a technological determinism: technology and whatever effects follow in its wake possess their own autonomous power, one that cannot be resisted or turned by individual or collective decisions. Proponents presume that CMC technologies will inevitably convey and reinforce specific cultural values – including preferences for free speech and individualism, especially as centralized control of information conveyed through the Internet and the Web is very difficult.

Implicating both technological instrumentalism and technological determinism, the image of an electronic global village thus rests on two fundamental but contradictory claims concerning technology. Beyond this underlying contradiction, both claims are themselves open to *empirically grounded* critique in light of what happens in praxis when CMC technologies,

---

[3] In addition to Ihde (1973, 1993), see Shrader-Frechette and Westra (1997) for an overview of these and other basic philosophies of technology.

[4] See Ess (1996) and Ess and Cavalier (1997) for an initial survey of the pertinent literature. The democratization theme is carried forward by numerous proponents of CMC technologies, including such well-known enthusiasts as Nicholas Negroponte (1995) and Bill Gates (1996). For more recent examples, see those documented by Harrison and Falvey (2001).

as initially designed almost exclusively within the cultural sphere of North America, are implemented and adopted in diverse cultural settings. Indeed, documented encounters between Western CMC technologies and diverse cultural groups demonstrate that *neither* technological instrumentalism *nor* technological determinism is consistent with what happens in praxis.[5]

Lorna Heaton (2001) describes some of the most striking examples of how Western-designed CMC technologies embody communicative preferences *not* shared by their intended users. In her study of Computer-Supported Collaborative Work (CSCW) systems, Heaton documents two Japanese *re*designs of CSCW systems that would more faithfully transmit the communication modes crucial within Japanese culture, which were thoroughly ignored in initial CSCW systems designed in the West. Heaton observes that much communication in Japan is nonlinguistic, involving body posture and distance, direction of gaze (including avoidance of eye contact), and hand gesture. In terms from communication theory, these nonverbal components are part of the Japanese *high context/low content* communication preference – a preference documented for other traditional and non-Western cultures as well (for example, Zaharna 1995). By contrast, Western – specifically North American white male – communication styles are *high content/low context.*

---

[5] Hofstede (1980) undertook a pioneering study of the intersections between culture and technology, resulting in a framework of five categories for evaluating cultural preferences, including the individualism/collectivism axis. The global quantitative study undertaken by Maitland and Bauer (2001) provides strong evidence that especially the axes of "masculine/feminine" (i.e., whether a culture endorses strong differences in gender roles rather than weak to no differences, respectively) and "low uncertainty avoidance" (i.e., openness to the risk associated with change), along with other cultural factors (English-language ability) and infrastructure factors, are correlated with more rapid technology diffusion within comparable countries/cultures. Indeed, such contrasts and correlations can now be marked out on an extensive global continuum, ranging from the relatively modest clashes between culture and technology apparent *within* the U.S. and Europe, through especially striking conflicts with the cultural values and communicative preferences of Middle Eastern, Asian, and indigenous cultures, to the conscious rejection of the Net in light of perceived cultural conflicts. As an example of the last point on the continuum, consider the South Pacific micro-state of the Kiribati. As described by Sofield (2000), the Kiribati see direct conflicts with the values embedded in CMC technologies and those of their own culture, namely: a commitment to economic equality that severely sanctions "shining" – i.e., standing apart from others in terms of material possessions, etc. – vs. the threat that unequal access to CMC technologies will issue in distinctive advantages for a few; a tradition of secrecy, especially with regard to governmental information, vs. the openness of CMC; and traditions of paternalism and communalism in government vs. the ways in which CMC technologies may foster individualism and individual independence. This last example may stand as a literally marginal case – as philosophers know, such margins and limits are essential, as they demarcate and define the boundaries of concepts, ideas, etc. In this instance, the Kiribati are a crucial example of the cultural demarcations of CMC technologies. In doing so, they further reinforce the central point: contra technological instrumentalism, CMC technologies are *not* culturally neutral. Rather, diverse cultural responses to these technologies – including their outright rejection – strongly argue that the technologies carry and favor specific cultural values and communicative preferences.

The standard ASCII e-mail exemplifies high content/low context communication. The *content* of the message constitutes most of the communication bandwidth: very little – if any – bandwidth indicates the relative social status of sender and receiver, their gender, and so forth. High content/low context communication is characteristic of Westerners, especially Americans, who value *direct* communication, *direct* eye contact, a presumed equality that ignores status differentials, and so on. But these same communication preferences, embedded in the very design of Western CSCW systems, were *not* comfortable for Japanese engineers, whose culture emphasizes instead *indirect* modes of communication made possible through nonverbal as well as verbal practices, for whom direct eye contact is often insulting, and so forth. In contrast with CSCW systems designed according to Western communicative preferences, Japanese CSCW systems capture the nonverbal modes of communication crucial to high context/low content communication. Rather than conveying relatively context-less "text-only" information, Japanese systems use high-bandwidth video systems to capture and convey communicators' body posture, distance, gaze, and hand gesture as crucial signs of social context.

These examples demonstrate that CMC technologies both embed specific cultural values and communicative preferences (contra technological *instrumentalism*), and (contra technological *determinism*) that these technologies do *not* simply reshape their users to conform with those embedded values and preferences. Rather, diverse peoples and cultures *are* capable of (re)designing systems more in keeping with their own cultural values and communicative preferences.

Several South Asian countries demonstrate similar patterns of response to Western-designed CMC technologies. For example, Abdat and Pervan (2000) analyze Group Support Systems in light of Indonesian cultural values. In terms familiar from Hofstede (1980), they characterize Indonesia as *low individualism/high collectivism* and *high power distance* (that is, endorsing large differentials between those at the top and those at the bottom of the social hierarchies of wealth, status, power, and so on). In this context, major meetings are not primarily occasions for egalitarian debate to achieve consensus. Rather, the details are already negotiated and planned in pre-meetings, almost always in deference to the opinions and views of those with the most prestige and power. The intention is to avoid surprises – especially those that lead to loss of face, a violation of a central value from Confucian ethics (Abdat and Pervan 2000, 211). Anonymity is touted in the West as a key advantage of CMC technologies, because it encourages more open and egalitarian communication, and may thus help flatten organizational hierarchies. Anonymity and equality, in short, are values of Western *high individualism/low collectivism* and *low power distance* cultures. But in the Indonesian context, anonymity is Janus-faced. On the one hand, in pre-meeting contexts where details are negotiated, anonymity *may* in fact contribute to group efficiency as it reduces status differentials.

On the other hand, in major meetings in which face is much more at stake, anonymity can encourage comments and questions that threaten face. For these reasons, Abdat and Pervan argue that GSS systems need to be redesigned in order to make anonymity a switchable feature – that is, so that these technologies cohere more closely with the cultural values and communicative preferences of their Indonesian users (2000, 213f.).

As a second example, Nasrin Rahmati (2000) compares Malaysian and Australian students using GSS systems. Rahmati characterizes Malaysian culture as marked by a distinctive religious-commitment factor, as well as high fatalism, high uncertainty avoidance, collectivism, traditionalism, and the value of keeping face. Such a society is denoted in sociological short-hand as a "tight" society. A "loose" society like Australia, by contrast, is marked by low religious commitment, low fatalism, low uncertainty avoidance, individualism, and so on. Rahmati finds that these cultural differences correlate with different responses to GSS technologies among Malaysian and Australian students. In particular, she finds that GSS tech-nologies "can change the social behavior of groups from different cultures and this change seems to be more pronounced for a sample from tight soci-eties than for a sample from loose societies" (2000, 271).

Rahmati's findings are consistent with those reported by Dahan (1999, 2001) and Wheeler (2001), and they make clear that CMC technologies can shift persons from their original cultural values and communicative prefer-ences. Both sets of findings, thus, counter technological instrumentalism. At the same time, however – contra technological determinism – this shift is neither absolute nor irresistible. On the contrary, especially Heaton's analy-sis makes clear that technology does not entirely remove us from our origi-nal cultural "skins" and somehow place us into some entirely new body or identity, as if we were Cartesian minds thoroughly disconnected from a body immersed in distinctive histories, traditions, and communities. (This point will be especially important for the discussion of *embodiment*, below.)

The Democratization Claim
These same results – a middle ground between technological instrumen-talism and technological determinism – can be seen in a specific example of presumed technological determinism: the claim that CMC technologies, as fostering more egalitarian communication, will inevitably result in democratization.

The democratization claim is now widely presumed and studied; in particular, it is the basis of any number of projects in the United States and elsewhere to establish "electronic democracy" via CMC technologies. The results from praxis, however, again provide grounds for both optimism *and* pessimism.[6] For example, Michael Dahan (1999) has argued that, *in*

---

[6] I cannot do justice here to the now exploding literature on electronic democracy. But it appears that while early enthusiasm for "electronic democracy" seemed justified by, most

conjunction with other social factors, CMC technologies have helped
Israel shift toward greater openness and democratic participation.
Indeed, in a remarkable experimental dialogue between Israelis and
Palestinians, Dahan successfully employed CMC technologies to overcome barriers between these bitter enemies, resulting in real-world, face-
to-face friendships – just before the escalation of violence in 2000
(Dahan 2001). But by itself, CMC does not *force* openness and participation to emerge. Similarly, Deborah Wheeler (2001) documents how
use of e-mail, Internet chatrooms, and the like, among women in Kuwait
can equalize communication between men and women in ways that
violate social traditions and expectations (such as the prohibition against
males and females speaking to one another in public). But this effect is
limited to younger women. Again, technology alone does not necessarily
force social change.

It is then not surprising to discover that in praxis CMC does not automatically generate democracy in the face of countervailing cultural factors
and conditions. So Sunny Yoon (2001) makes clear that, contrary to hopes
for establishing an especially Habermasian public sphere in cyberspace
(cf. Becker and Wehner 2001), *unequal* access and distribution of CMC
technologies are fostered by commercialization and Korean journalism's
highlighting eCommerce as the most prominent and interesting use of
these technologies. Yoon also observes the importance of English as a
"cultural capital" (a concept developed by Bourdieu), a capital that
involves a subtle but nonetheless intransigent sort of power (Yoon 2001,
245, 256–58). These results in Korea reinforce Maitland and Bauer's finding that English-language ability is a significant cultural factor in the diffusion of CMC technologies (Yoon 2001).

Yoon's analysis again shows that if CMC technologies do indeed embed
Western preferences for open communication, equality, and so forth, these
technologies by no means inexorably impose these preferences on people
shaped by different cultural values. Thus, CMC technologies are ambiguous: as Yoon emphasizes, these technologies may lead *either* to greater *or*
less democracy and equality, depending on social and individual choices –
that is, on the *social context of use*. A number of additional important studies make this same point, perhaps most notably Soraj Hongladarom's
analysis (2001) of how Thai participants in a USENET newsgroup manage
to take up Western-based CMC technologies in ways that provide for both
global connectivity (in what Hongladarom characterizes as "thin" culture,

famously, the PEN experiments of the late 1980s and early 1990s, these experiments are
now seen to have failed in important ways, and optimism regarding the power of CMC to
encourage democratic forms of communication and polity is now much more guarded. In
particular, consistent with the recent turn in scholarship that I describe in the conclusion
below, there is a much greater awareness of precisely the need to pay attention to the social
context of use – to the larger social institutions and values that surround the effort to implement electronic democracy (e.g., Miani 2000; Harrison and Falvey 2001).

following Michael Walzer) and the preservation and enhancement of local (or "thick") culture.

In sum, these empirically grounded analyses of what happens in praxis when Western CMC technologies are deployed in diverse cultural settings allow us to test the twin philosophical claims of technological instrumentalism and technological determinism. So far, at least, the sorts of data collected in these analyses provide strong *empirical* grounds for rejecting *both* claims. Neither claim is consistent with the results of implementing CMC technologies in praxis. Contra technological instrumentalism, CMC technologies are not value neutral but rather embed and foster the cultural values and communicative preferences of their designers. At the same time, however, these technologies cannot simply overrun "target" cultures and homogenize them into a (Western) cultural monolith: by themselves, CMC technologies are *not* sufficient for an inexorable process of democratization – or, more broadly, computer-mediated colonization. Rather – and contra the claims of technological determinism – both democratization in particular and larger potential social impacts in general are tempered, sometimes profoundly, by the values and communicative preferences of those who take up the new technologies.

Philosophically, these results point us to a middle ground between technological instrumentalism and technological determinism. Such a middle ground has been demarcated as *soft determinism* by philosopher of technology Don Ihde (1973).[7]

Metaphysics/Epistemology

On the Self

Parallel to the cross-cultural encounters from the Ionians and the Milesians through the Renaissance and modernity, the manifold and diverse cultural encounters through CMC have made possible – even required – a careful examination of our most basic worldview assumptions. Beyond illuminating such issues as technological determinism and instrumentalism, these cross-cultural encounters further provide insight – perhaps distinctively new insight – concerning the nature of the self and epistemology.

The computer-mediated conversation about the self begins, of course, in numerous postmodern approaches to CMC, perhaps best represented by Lyotard (1984), Bolter (1984, 1991, 2001), and Landow (1992, 1994), who argue that the fluid and hypertextual forms of communication facilitated by computer networks and the Web closely resemble and realize postmod-

---

[7] Philosophers of technology have criticized technological determinism on a number of grounds. For a review of the pertinent literature in philosophy as well as communication theory, see Ess 2000. In addition, see Heaton for her discussion and documentation of "a growing backlash against technological determinism, [and] an increasing awareness that the path a given technology takes may not be inevitable and absolute," in social-science approaches to technology and engineering (2001, 215).

ern accounts of the self as fluid, fragmentary, decentered, and so on. This postmodern thread of discussion, however, is countered by more recent turns in the literatures of CMC and related fields that take, especially, phenomenological and hermeneutical approaches to argue for the self as tied closely to *embodiment*.[8]

The cultural encounters documented here helpfully illuminate these analyses. For example, Heaton's account of Japanese redesign of CSCW systems brings to the foreground the significance of nonverbal communication in Japanese culture and the resulting contrast between Japanese high context/low content preferences and Western high content/low context preferences. This Western high content/low context preference, moreover, is consistent with a prevailing Cartesian dualism, one that minimizes the role of body in knowing and communicating, while maximizing reason and rational discourse as a locus for knowledge and communication both separate from and superior to body. By contrast, Japanese emphasis on high context/low content communication – shared by other Asian and traditional cultures – is consistent with a nondualistic, non-Cartesian sense of self.[9]

Similar remarks can be made with regard to other elements of communicative analysis. For example, Hofstede's contrast between low individualism/high collectivism societies vis-à-vis high individualism/low

[8] Somewhat more carefully, a specifically philosophical discussion of the nature of the self as illuminated by *interaction* with computers is at work in the earliest stages of artificial intelligence and the debates, for example, between nondualistic views and more Cartesian views that emphasize the radical split between body and reason/mind as the intelligence to be replicated by the computer, represented in the pioneering work of Douglas Engelbart (see Bardini 2000) and Winograd and Flores (1987). See Ess 2002, for a more complete discussion.

[9] While philosophers have been aware of the role of culture in shaping worldview, including our sense of self, of who we are as human beings, since at least pre-Socratics like Xenophanes (horses would have gods resembling horses, fragment 15; see Kirk and Raven 1966, 169), to my knowledge, philosophers have not examined the role of culturally shaped *communication* preferences as consistent with and perhaps contributing to culturally shaped notions of self. This approach is similar to lines of inquiry regarding the role of language in shaping worldview developed by Whorf, Wittgenstein, and the "ethno-metaphysics" of Overholt and Callicott (1982). See as well Ames and Rosemont 1998, 20–35, for a careful analysis of the nondual sense of self that correlates with the structure of classical Chinese. It differs, however, as it conjoins Hofstede's (and subsequent) analyses of culture and technology and an important communication theory developed by H. A. Innis, Elizabeth Eisenstein (1983), Marshall McLuhan (1965), and Walter Ong (1981, 1988) that develop contrasts between orality, literacy, and the "secondary orality of electronic culture" (Ong 1988, 135–38) as important stages of cultural development, stages defined (if not determined) by these basic forms of communication taken to operate as a *technology*. Chesebro and Bertelson (1996) provide a very useful overview of this theory. At the same time, however, it must be emphasized that this schema, as it makes a strong distinction between orality, literacy, and the "secondary orality of electronic culture," has been criticized on a number of grounds by cultural studies and communication theorists as too simple and too deterministic. See Sveningsson (2001) for an especially helpful overview of critiques from communication theoretical perspectives.

collectivism societies (noted above in Abdat and Pervan's analysis of Indonesia) provides another culturally dependent lens through which to consider philosophical claims of the self. As the Japanese example suggests, Asian societies stressing low individualism/high collectivism show a marked contrast to a Western concept of the self as an individual *apart from* the community (beginning with Hobbes's atomistic conception of self through Descartes's radical dualism). These philosophical conceptions of self, especially ones affiliated with Cartesian dualism (self as mind divorced from body, self as individual apart from community) can hence be called into question not only on philosophical grounds but also by appeal to the empirical accounts of communication, language, technologies of communication (including orality, literacy, print, and what Ong calls "the secondary orality" of electronic culture), and correlative notions of self that come to the foreground through examining the implementation of Western CMC systems in praxis.

As a last example, the various ways in which we can see that individuals and cultures both adapt to and resist Western-based CMC technologies – most notably, Hongladarom's account of Thais as engaged in both thin culture and thick culture – provides another layer of empirically grounded evidence regarding the self. This account suggests that the self is *both* deeply shaped by and anchored in a specific culture – including a specific range of communication preferences – *and* capable of developing new modes of communication and participating in "different" sorts of cultures.

The suggestion of a (non-Cartesian) self that remains rooted (via) body in a thick local culture (including its language, history, and traditions) while simultaneously engaged in thin but global Internet culture, may help resolve an emerging *political* problem occasioned by CMC technologies. Of course, there have long been "multicultural persons" (Adler 1977) who have learned how to be at home in more than one culture. But a crucial question raised by the explosive emergence of a global communications web is whether these forms of CMC will in fact facilitate a move by significant numbers of their users to a genuinely more cosmopolitan and *intercultural* global village – and/or whether, contrary to proponents' claims, these technologies will only reify the cultural values and communicative preferences of the already dominant Western culture, thus serving as a form of computer-mediated colonization that will homogenize world cultures (Ess and Sudweeks 2001, 259f.). It is far too early in the development of global CMC technologies even to hazard a guess. But the Thai example and others (such as Thanasankit and Corbitt 2000) suggest that at least (non-Cartesian) multicultural persons can create the social contexts of use for these technologies that will lead to implementations of CMC of global scope that will also preserve and enhance local cultural identities, avoiding an electronic "colonization of the lifeworld" (so Peter Sy 2001, 305–08) – a colonization threatened by the fact that CMC technologies *do* embed specific cultural values and communicative preferences that, as the example of Korea makes clear, can be further shaped by commercialization.

Epistemological Gleanings

Intimately interwoven with questions regarding the self, of course, are questions of epistemology. We have already seen that some of the results of implementing CMC technologies in diverse cultural settings call into question an especially Cartesian view of the self as a rational mind radically divorced from body. But of course this is at once an *epistemological* point: is knowledge located solely in such a Cartesian mind divorced from body, or is knowledge distributed across multiple sites of the self, as a continuum of mind and body, what Barbara Becker (2000) calls the "body-subject"?

As we have already seen,[10] these epistemological questions are at stake in the earliest stages of artificial intelligence, and in the debates between more Cartesian views, which emphasize the radical split between body and reason or mind as the intelligence to be replicated by the computer, and nondualistic views, represented in the pioneering work of Douglas Engelbart (see Bardini 2000) and Winograd and Flores (1987). While Cartesian dualism is further apparent in the feminist manifesto of the early Donna Haraway (1990) and the "ectopian" hopes of Hans Moravec (1988), it is increasingly supplanted by epistemologies and senses of self marked by the inextricable *connection* between body and mind. So, for example, Katherine Hayles characterizes the "posthuman" in terms of a specific *epistemological* agenda: "Reflexive epistemology replaces objectivism . . . embodiment replaces a body seen as a support system for the mind; and a dynamic partnership between humans and intelligent machines replaces the liberal humanist subject's manifest destiny to dominate and control nature" (Hayles 1999, 288). Hayles thus shifts from an objectivist epistemology (resting on a dualistic separation between subject and object, and thus between subjective vs. objective modes of knowledge, coupled with the insistence that only "objective" modes of knowledge are of value) to an epistemology that (echoing Kant) stresses the inextricable interaction between subject and object in shaping our knowledge of the world. In the same way, Hayles further focuses precisely on the meanings of *embodiment* in what many now see as a post-Cartesian understanding of mind-*and*-body in cyberspace.[11]

These debates are further illuminated by specific implementations of

---

[10] See note 8, above.

[11] In addition to Becker (2000, 2001), see Bolter 2001; Brown and Duguid 2000; Dertouzos 2001. Again, this nondualistic conception of self coheres with the classical Chinese notion of *xin*, "heart-and-mind" as translated by Ames and Rosemont (1998, 56), and the process-oriented character of classical Chinese (1998, 20–35). I explore this connection between a contemporary Western turn from Cartesian dualism to nondualistic conceptions of self more fully below, in note 16.

CMC technologies among indigenous peoples. For example, Louise Postma (2001) has documented the ways in which learning centers in South Africa, including computer-based resources, embody a distinctive epistemology: not surprisingly, these centers foster the epistemological assumptions and orientation of their designers' white European cultures. *Learning* in this context focuses on the (Cartesian) *individual* acquisition of knowledge through *texts* in an atmosphere of quiet that prohibits social discourse. By contrast, the preferred (non-Cartesian) learning styles of the indigenous African peoples Postma describes are *collaborative* and grounded in *orality* and *performance*: most simply, "knowledge" is gained through oral communication, repetition, and the performance of important social narratives. If CMC technologies are to fulfill their promise for greater equality and what Friere calls *critical empowerment* – that is, the preservation and enhancement of a community's specific epistemologies and communicative preferences – these technologies must be designed and used in ways that are consistent with the cultural preferences for orally based, collaborative approaches to gaining knowledge. More broadly, the contrast between Western Cartesian and indigenous non-Cartesian episte-mologies and learning styles brought to the foreground in these and simi-lar examples (see Harris et al. 2001 and Sy 2001) again suggests from an empirically grounded viewpoint that the Cartesian epistemology and correlative sense of individual sense is not universal but culturally limited.[12]

Concluding Remarks

From the multiple interactions between Western-based CMC technologies and diverse global cultures, philosophers may thus glean insights from praxis that shed light on debates in philosophy of technology and the perennial philosophical inquiries into knowledge and the nature of the self. By way of conclusion, let me point to additional aspects of CMC technol-ogy that likewise promise new insights for philosophers willing to under-take a new, praxis-oriented, interdisciplinary approach to their favorite questions and problems.

To begin with, these sorts of analyses of CMC technologies are largely limited to descriptions of what happens *after* implementing a given CMC technology. With few exceptions (Turk and Trees 1998; Harris et al. 2001), these analyses and projects do *not* include careful analysis of cultural values for purposes of shaping the design and implementation of CMC

---

[12] Indeed, there is a remarkable resonance here between the collaborative/kinesthetic epistemologies of orality and Becker's phenomenological account of the "bodysubject": "Rhythm and sound, proximity and touch, gesture, mimicry and movement: the body becomes voice [*meldet sich . . . zu Wort*] in the most diverse ways and thus also colors its language – not only the spoken language (through tone, turn of phrase, and sound) but also the written language (e.g., in poetry and metaphor)." (2001)

technologies. But the exceptions make clear that the design and implementation of CMC technologies in diverse cultural settings may be intentionally shaped and controlled so as to test and refine specific hypotheses in the light of the empirical results. These more active philosophical engagements with CMC allow philosophers to approximate more closely the model of the natural-science laboratory.[13]

Mike Sandbothe suggests an example of such an active approach. As he points out, persons who participate in MUDs or MOOs – that is, primarily text-based forms of interaction online – are thereby forced explicitly to articulate their identity *in language* to others in ways that remain tacit in ordinary face-to-face communication. These senses of identity include the most basic elements of worldview, including assumptions concerning the singularity or multiplicity of identity (as we present ourselves to others under a single pseudonym and/or multiple pseudonyms and avatars – a sort of deception easily possible, even encouraged, in the pseudonymous or anonymous text spaces of a MUD or MOO, but far more difficult for embodied creatures in a face-to-face world). Also, these contexts enable almost instant communication across a globe of markedly different, real-world "times" as well as asynchronous communications that can suspend the ontological and ethical importance of a shared *time*. In this way, our received notions of time and space are suspended, made articulate, and open – up to a point, at least – to our reshaping time and space according to our own preferences and needs (1999, 430f.).[14]

In this fashion, CMC technologies provide new environments that require us to engage in the classically philosophical project of bringing to the foreground and then subjecting to critical scrutiny our own most basic assumptions, including our assumptions concerning time and space. And insofar as CMC provides environments in which we can construct new identities and virtual worlds based on alternative conceptions of space, time, relationship, and so forth, they would allow us to test these as hypotheses from a variety of philosophical standpoints. For example,

---

[13] This is *not* to suggest a slavishly positivist approach to philosophy that evaluates the legitimacy of philosophical claims in the light of the presumably superior paradigm of scientific procedure. On the contrary, it is an effort to overcome the arguably artificial divide between philosophy and the natural sciences – in part, by understanding the empirical orientation of the natural sciences as a specific instance of the larger Aristotelian/Habermasian emphasis on the importance of testing philosophical claims by way of implementing them in praxis.

[14] For phenomenological counterpoints to the (postmodern) enthusiasm represented here by Sandbothe for a postmodern self set free in cyberspace, see Kaltenborn 2001 and Meyer-Drawe 2001. Kaltenborn and Meyer-Drawe articulate phenomenologically grounded aspects of *embodiment* that become lost in cyberspace and thereby stand as additional examples of the recent turn from postmodern/Cartesian dualism to nondualistic conceptions of self as embodied, a turn we have seen exemplified in the work of Hayles (1999) and Becker (2000). Similarly, Hillis (1999) critiques Virtual Reality as resting on postmodern/Cartesian dualisms, one resulting precisely in the sort of schizophrenia that Kaltenborn discusses (see especially Hillis, ch. 6, "Identity, Embodiment, and Place: VR as Postmodern Technology").

much has been made of online or virtual communities, initially as replacements for real-world, face-to-face communities and relationships. In parallel with the turn noted above and exemplified in Hayles's notion of the postmodern human, recent research suggests that exclusively online relationships tend to fade over time; but online relationships may be used successfully to develop and sustain real-world relationships. From these data, a philosopher could propose various forms of possible conjunctions of real-world and virtual identities and communities, and analyze their success in establishing and sustaining human relationships and communities on phenomenological, ethical, and/or pragmatic grounds.

Indeed, such experiments may prove to provide crucial philosophical insights that might help resolve significant real-world problems. For Sandbothe, as CMC requires us to become more philosophically explicit and intentional regarding our most basic assumptions concerning identity, time, and space, we may come to see the *contingent* and *non*-universal character of these assumptions. In my terms, CMC may thus result in an *epistemological humility* – one that, to continue with Sandbothe's argument, would support Richard Rorty's hope that the new media will lead to a transcultural communication. As CMC technologies facilitate the distinctively philosophical project of making our basic assumptions more articulate, and thereby encouraging a posture of epistemological humility as we recognize that our most basic beliefs enjoy only limited certainty and universality, we may likewise become more empathic, understanding, and receptive toward the "others" of different identities and cultures, the others with whom we can now communicate in the extraordinary communicative flows made possible by these same technologies. We have seen for example, how CMC technologies as implemented among non-Western and indigenous peoples suggest that a Cartesian conception of self and epistemology is distinctively Western and culturally limited. Western users of these technologies should recognize these cultural limitations concerning our most fundamental assumptions and, in an epistemological humility that makes us more conscious of these limitations, take care not to assume – or inadvertently impose – distinctively Western presumptions regarding the nature of the self in communicating with others.

In other terms, CMC technologies offer philosophers new environments that, following Plato's well-known allegory, may facilitate escaping the chains of worldviews sustained only by cultural authority and tradition in the cave, to discover or invent new, more comprehensive worldviews – ones that can be articulated and tested in the electronic spaces of CMC. As Dahan's remarkable experiment with CMC and Habermasian rules of discourse to facilitate communication between Israelis and Palestinians has demonstrated, CMC technologies *can* facilitate extraordinary new conversations between and among even those cultures initially marked by the greatest antipathy toward one another (Dahan 2001). More broadly, this becomes a testable hypothesis – one calling for any number of experiments to determine whether, and if so, how and under what conditions, CMC technologies can regularly lead to such philosophical self-aware-ness, epistemological humility, and greater cross-cultural understanding.

Interdisciplinarity, Philosophy of Communication, Global Ethics, and Socratic-Renaissance Education

Western philosophy arises precisely from the encounter with diverse cultures and peoples as it emerges out of the Ionian and Milesian colonies. In the face of the conflicting religions and mythologies of many peoples, the Pre-Socratics begin the distinctively philosophical task of uncovering and critically examining underlying assumptions from the standpoint of *logos* ("reason") rather than *mythos*. As becomes explicit in Aristotle – and later in Kant and Hegel – they thereby develop an alternative to myth in the form of a complete logos (account or explanation) of the world (human, natural, and divine) as an intelligible place ordered by logos (principle, and so forth). Other periods of heightened cultural flows have similarly witnessed new blooms of philosophical insight and creativity – for example, the emergence of what becomes modern natural science out of the interactions between Jewish, Christian, Islamic, and Chinese sources in the Middle Ages.

The sorts of cultural engagements in our time made possible by the Internet and the Web both parallel and dramatically expand the cultural flows surrounding the emergence of Western philosophy and some of its most creative developments. Beyond their service as a kind of cultural and communicative laboratory, by facilitating cultural flows and engagements on a new scale and scope, CMC technologies promise to facilitate creative developments in philosophy analogous to the emergence of modern natural science and, indeed, the emergence of Western philosophy itself. For example, an intercultural dialogue among philosophers representing a variety of traditions (for example, from Western Aristotelian virtue ethics to Confucian, Taoist, Buddhist, and Hindu thought) may find new commonalities and approaches for a global ethics.[15] Movement in this direction, of course, would realize the ancient Stoic dream of the cosmopolitan, the citizen of the world who is not simply defined by the worldview of a specific culture (thick culture, to borrow from Hongladarom) but who also participates in a global (if thin) culture. Perhaps most grandly, hypertext and multimedia may facilitate an even more comprehensive interdisciplinary dialogue, one realizing new forms of knowledge anticipated by Nietzsche's conception of the "gay science," the *fröhliche Wissenschaft* that reconnects the seriousness of the natural and humanistic sciences (*Naturwissenschaften* and *Geisteswissenschaften*,

---

[15] In addition to the multiple publications by Bynum and Moor (referenced elsewhere in this collection), see, as but representative examples, Adler 1977; Boss 1998a, 1998b; Blocker 1999; Gupta and Mohanty 2000; Mall 2000, as well as the publications of the Académie du Midi, e.g., Elberfeld et al. 1998.

respectively, the latter including philosophy) with the creativity and play-fulness of poetry (Kaufmann 1974, 4–13).

Such possibilities, however, will require interested philosophers to turn to new ways of doing philosophy – first of all, in the interdisciplinary approaches required as philosophers move beyond their own disciplines and subdisciplines to collaborate with the range of researchers and scholars who help us better understand CMC and culture.

In part, such interdisciplinary approaches are already at work in the larger context of *philosophy of communication*, as represented in such philosophers as Wittgenstein, Austin, and Searle and most especially in the communicative models of rationality that emerge from Gadamer, Ricoeur, Habermas, Apel, Bahktin, Davidson, and Rorty. In particular, the empirical work surveyed here highlights cultural contrasts in communication preferences, including the relative importance of verbal and nonverbal communication and high content/low context vs. high context vs. low content communication. This work further foregrounds the relationship between different media (orality, literacy, print, and the secondary orality of cyberspace) and ways of thinking about self and world, what counts as knowledge, and so on. As philosophers of communication, we need to continue to rethink our assumptions concerning what counts as the "communication" and dialogue central to the philosophical enterprise, and thereby what *philosophical claims, evidence, and arguments* can and cannot be conveyed via print and text. To put it crudely: Are our current conceptions of philosophical methods and claims delimited to an argumentative space that attends only to high content/low context verbal and textual modes of communication, in the service of a solitary Cartesian knower as the exclusive model of rationality, thereby excluding other forms of knowledge claims, including the nonverbal, the kinesthetic, the collaborative, and all else associated with high context/low content communication? More constructively: out of the interdisciplinary approaches of philosophy of communication, this essay sketches out a trajectory toward a more inclusive understanding of communication and less ethnocentric conception of philosophy. It relies on communicative, cultural, and social scientific approaches to CMC to support a move beyond Cartesian dualism to more holistic metaphysics and epistemologies. These more holistic metaphysics and epistemologies, finally, point toward a correlative, genuinely *global* ethics – one that could, for example, involve a coalescence between Aristotelian and Confucian ethics.[16]

---

[16] In Confucian ethics the term used to describe personhood, *ren*, sharply contrasts with a Cartesian dualism, as "*Ren* is not only mental, but physical as well: one's posture and comportment, gestures and bodily communication" (Ames and Rosemont 1998, 49). Similarly, Ames and Rosemont render *xin* as "heart-and-mind," contra the Western mind-body dichotomy, to make the point that in Confucian thought one never finds either pure thought devoid of feeling or pure feeling devoid of thought (1998, 56). These classical Chinese conceptions thus cohere with contemporary Western efforts to overcome a radical

In particular, I have argued more fully elsewhere that in addition to developing a global ethics philosophers can contribute to a new form of education, shaped by our best understanding of *what it means to be human* in these new environments and how one best communicates cross-culturally in these new environments (Ess 2000, 2002). Parallel to the arguments of Cees Hamelink (2000), this education will be in part a *Socratic* one that will parallel the Renaissance as it seeks to help human beings become "cultural polybrids" who are fluent in and comfortable with the languages, customs, and worldviews of multiple cultures.[17]

But while the technologies that occasion and facilitate these philosophical turns are new, the turns themselves are also *returns*. That is, philosophy as practiced in interdisciplinary dialogue with others, and oriented toward shaping both the praxis of CMC technologies and the Socratic education of their users, goes beyond the current boundaries of philosophy as a technical discipline – boundaries shaped less by the projects of philosophy itself and more by positivism and the industrialization of university education over the past century or so. Philosophers engaged in broadly interdisciplinary dialogues and projects may thereby recover something of the original impetus of Western philosophy to develop a comprehensive logos – however much this project needs refinement in light of feminist and postmodernist critiques of at least *modern* forms of rationality – and they may perhaps contribute to the emergence of new forms of knowledge (echoing the Middles Ages and the Renaissance). At the same time, such a return may recover the original insistence on praxis, including the ethical and the political. In particular, revitalizing a Socratic education could help resolve the political problem of cultural homogenization in the form of computer-mediated colonization, as this education aims at helping human beings move beyond – but not abandon – the cave of specific cultural worldviews for the sake of becoming the Renaissance women and men who can utilize

mind-body split – including, as we have seen, Hayles's notion of a post-Cartesian self and Becker's "bodysubject" (2000). Moreover, both Socratic and Aristotelian ethics emphasize the primary importance of developing those habits (*ethos*) that allow us to become "virtuous" (meaning *excellent*) human beings (Aristotle, *Nichomachean Ethics*, book 2, 1103a 14–26). Similarly, Confucian ethics emphasizes the life-goal of becoming a *junzi*, an exemplary person (Ames and Rosemont 1998, 62f.).

[17] The Renaissance model educates human beings in the languages, values, beliefs, and practices of multiple cultures, thus moving us beyond our own culture to inhabit the life-worlds of genuinely different cultures and peoples. A Socratic education emphasizes critical thinking and dialogue – but also a deep engagement with "other" cultures, languages, and worldviews as a way of moving out of our particular cultural cave to a respectful appreciation of diverse cultures and a correlative epistemological *humility* regarding any single claim or worldview. In addition, a Socratic-Renaissance form of education fulfills Habermasian and feminist requirements for perspective taking and solidarity. Especially as conjoined with, say, Confucian ethics, such an education takes seriously the role of living and knowing as an embodied creature ("bodysubject"/*xin*) in a world of multiple cultures and peoples, and of moving in more just directions. It may serve as a more genuinely *global* ethics, required for the Internet and the Web as global technologies. See Ess 2000, 2002.

these new communication technologies in a praxis and ethos that instead preserve and enhance diverse cultures.[18]

References

Adler, Peter S. 1977. "Beyond Cultural Identity: Reflections on Multiculturalism." In *Culture Learning: Concepts, Applications, and Research*, edited by Richard W. Brislin, 24–41. Honolulu: University of Hawaii Press.

Ames, Roger, and Henry Rosemont Jr. 1998. *The Analects of Confucius: A Philosophical Translation*. New York: Ballantine Books.

Bardini, T. 2000. *Bootstrapping: Douglas Engelbart, Coevolution, and the Origins of Personal Computing*. Stanford: Stanford University Press.

Becker, Barbara. 2000. "Cyborg, Agents and Transhumanists." *Leonardo* 33, no. 5: 361–65.

———. 2001. "Sinn und Sinnlichkeit: Anmerkungen zur Eigendynamik und Fremdheit des eigenen Leibes" [Sense and sensibility: Remarks on

---

[18] This emphasis on a *continuity* between a contemporary global computer ethics and such moments in the history of Western ideas as the emergence of philosophy, the Middle Ages, and the Renaissance, sharply contrasts with Gorniak-Kocikowska's insistence on a radical diremption between any future global computer ethics and earlier ethical systems (1996). Gorniak-Kocikowska's view of computer ethics rests in part on claims inconsistent with more recent theoretical and empirical analyses. First, Gorniak-Kocikowska makes the common (especially postmodern) claim that the computer revolution is as radical a revolution as the invention of the printing press (1996, 181f.), a view, as we have seen, that has been rejected on several grounds (e.g., note 9, above). Gorniak-Kocikowska further assumes a strongly *dualistic* relationship between new and previous technologies, and thereby between the global ethics facilitated and required by new information technologies and previous ethical systems (specifically, Kant and Bentham – see 1996, 188). Underlying this dualism is a presumption of *technological determinism*, so she argues that "the future global ethic will be a computer ethic because it will be caused by the Computer Revolution and will serve the humanity of a Computer Era" (179). But we have seen that technological determinism is rejected in philosophy of technology and communication theory – as well as by our observations of what happens in praxis as CMC technologies are deployed in diverse cultural settings. Moreover, Gorniak-Kocikowska's dualistic either/or between contemporary and earlier technologies and their affiliated ethics leads to a particular claim – namely, that "actions in cyberspace won't be local. Therefore, the ethical rules for these actions cannot be rooted in a particular culture" unless those ethical rules are applied homogenously (imperialistically) across the globe – an unacceptable alternative, of course (188). By contrast, I have tried to argue that more recent literature and insights in CMC emphasize instead a "both/and" logic of *connectedness* and interconnection, most evident in Soraj Hongladarom's model for a global Internet: thin culture *in conjunction with* the values and preferences of thick local cultures. Rather than the either/or Gorniak-Kocikowska presents between contemporary and earlier technologies, and thus between a global computer ethics and the ethical frameworks of diverse cultural traditions, Hongladarom's model suggests the possibility of a thin global computer ethics (as I have suggested here, one that includes a focus on *virtue* as understood in both Western and Eastern thought) but further includes thick local traditions like (Western) Kantianism and utilitarianism *and*, say, (Eastern) Confucian ethics.

238 CHARLES ESS

the distinctive dynamics and strangeness of one's own body]. In *Mentalität und Medialität*, edited by L. Jäger. Munich: Fink Verlag.

————, and J. Wehner. 2001. "Electronic Networks and Civil Society: Reflections on Structural Changes in the Public Sphere." In *Culture, Technology, Communication: Towards an Intercultural Global Village*, edited by Charles Ess, 65–85. Albany, N.Y.: State University of New York Press.

Blocker, H. Gene. 1999. *World Philosophy: An East-West Comparative Introduction to Philosophy*. Upper Saddle River, N.J.: Prentice-Hall.

Bolter, J. D. 1984. *Turing's Man: Western Culture in the Computer Age*. Chapel Hill: University of North Carolina Press.

————. 1991. *Writing Space: The Computer, Hypertext, and the History Of Writing*. Hillsdale, N.J.: Lawrence Erlbaum.

————. 2001. "Identity." In *Unspun*, edited by T. Swiss, 17–29. New York: New York University Press. Available online at <http://www.nyupress.nyu.edu/unspun/samplechap.html>.

Borgmann, A. 1984. *Technology and the Character of Contemporary Life*. Chicago: University of Chicago Press.

————. 1999. *Holding onto Reality: The Nature of Information at the Turn of the Millennium*. Chicago: University of Chicago Press.

Boss, Judith. 1998a. *Ethics for Life: An Interdisciplinary and Multicultural Introduction*. Mountain View, Calif.: Mayfield.

————. 1998b. *Perspectives on Ethics*. Mountain View, Calif.: Mayfield.

Brown, J. S. and P. Duguid. 2000. *The Social Life of Information*. Stanford: Stanford University Press.

Chesebro, James, and Dale A. Bertelsen. 1996. *Analyzing Media: Communication Technologies as Symbolic and Cognitive Systems*. New York: Guilford Press.

Dahan, M. 1999. "National Security and Democracy on the Internet in Israel." *Javnost the Public*, 6, no. 4: 67–77.

————. 2001. Personal communication.

Dertouzos, M. 2001. *The Unfinished Revolution: Human-Centered Computers and What They Can Do for Us*. New York: HarperCollins.

Eisenstein, Elizabeth. 1983. *The Printing Revolution in Early Modern Europe*. Cambridge: Cambridge University Press.

Elberfeld, Rolf, Johann Kreuzer, John Minford, and Günter Wohfart, eds. 1998. *Komparative Philosophie: Begegnungen zwischen östlichen und westlichen Denkwegen*. Schriften der Académie du Midi, volume 4. Munich: Wilhelm Fink Verlag.

Ess, Charles. 1996. "The Political Computer: Democracy, CMC, and Habermas." In *Philosophical Perspectives on Computer-Mediated Communication*, edited by Charles Ess, 197–230. Albany, N.Y.: State University of New York Press.

————. 2000. "We Are the Borg: The Web as Agent of Cultural

Assimilation or Renaissance?" Available online at <www.ephilosopher. com/120100/philtech/philtech.htm>.

————. 2002. "Computer-Mediated Communication and Human-Computer Interactions." In *Blackwell's Guide to the Philosophy of Computing and Information*, edited by Luciano Floridi. Oxford: Blackwell.

Ess, Charles, and Robert Cavalier. 1997. "Is There Hope for Democracy in Cyberspace?" In *Technology and Democracy: User Involvement in Information Technology*, edited by David Hakken and Knut Haukelid, 93–111. Oslo, Norway: Center for Technology and Culture.

Ess, Charles, and Fay Sudweeks. 2001. "On the Edge: Cultural Barriers and Catalysts to IT Diffusion among Remote and Marginalized Communities." *New Media and Society* 3, no. 3: 259–69.

Gates, Bill, with Nathan Myhrvold and Peter Rinearson. 1996. *The Road Ahead*. New York: Penguin Books.

Gorniak-Kocikowska, Krystyna. 1996. "The Computer Revolution and the Problem of Global Ethics." *Science and Engineering Ethics* 2: 177–90.

Grim, Patrick, Trina Kokoalis, Ali Alai-Tafti, Nicholas Kilb, and Paul St. Denis. 2001. "The Emergence of Communication: Some Models for Meaning." Sixteenth Annual Computers and Philosophy Conference, Carnegie Mellon University, Pittsburgh. 10 August.

Grim, Patrick, Gary Mar, and Paul St. Denis with the Group for Logic and Formal Semantics. 1999. *The Philosophical Computer: Exploratory Essays in Philosophical Computer Modeling*. Cambridge, Mass.: MIT Press/Bradford Books.

Gupta, Bina, and J. N. Mohanty, eds. 2000. *Philosophical Questions East and West*. Lanham, Md.: Rowman and Littlefield.

Hall, Kira. 1996. "Cyberfeminism." In *Computer-Mediated Communication: Linguistic, Social, and Cross-Cultural Perspectives*, edited by Susan Herring, 147–70. Amsterdam: John Benjamins.

Hamelink, Cees. 2000. *The Ethics of Cyberspace*. London: Sage Publications.

Haraway, Donna J. 1990. "A Cyborg Manifesto: Science, Technology, and Socialist-Feminism in the Late Twentieth Century." In *Simians, Cyborgs, and Women: The Reinvention of Nature*, 149–81. New York: Routledge.

Harris, Roger, Poline Bala, Peter Songan, Elaine Khoo Guat Lien, and Tingang Trang. 2001. "Challenges and Opportunities in Introducing Information and Communication Technologies to the Kelabit Community of North Central Borneo." *New Media and Society* 3, no. 3: 271–96.

Harrison, Teresa M., and Lisa Falvey. 2001. "Democracy and New Communication Technologies." In *Communication Yearbook* 25, edited by William B. Gudykunst. Hillsdale, N.J.: Lawrence Erlbaum.

Hayles, K. 1999. *How We Became Posthuman: Virtual Bodies in Cybernetics, Literature, and Informatics*. Chicago: University of Chicago Press.

Heaton, L. 2001. "Preserving Communication Context: Virtual Workspace and Interpersonal Space in Japanese CSCW." In *Culture, Technology, Communication: Towards an Intercultural Global Village*, edited by Charles Ess, 213–40. Albany, N.Y.: State University of New York Press.

Herring, Susan. 1996. "Posting in a Different Voice: Gender and Ethics in Computer-Mediated Communication." In *Philosophical Perspectives on Computer-Mediated Communication*, edited by Charles Ess, 115–45. Albany, N.Y.: State University of New York Press.

————. 1999. "The Rhetorical Dynamics of Gender Harassment On-line." *The Information Society*, 15, no. 3: 151–67.

Hillis, Ken. 1999. *Digital Sensations: Space, Identity and Embodiment in Virtual Reality*. Minneapolis and London: University of Minnesota Press.

Hofstede, Gert. 1980. *Culture's Consequences: International Differences in Work-Related Values*. London: Sage Publications.

Hongladarom, S. 2001. "Global Culture, Local Cultures, and the Internet: The Thai Example." In *Culture, Technology, Communication: Towards an Intercultural Global Village*, edited by Charles Ess, 307–24. Albany, N.Y.: State University of New York Press..

Ihde, Don. 1973. "A Phenomenology of Man-Machine Relations." In *Work, Technology, and Education: Dissenting Essays in the Intellectual Foundations of American Education*, edited by Walter Feinberg and Henry Rosemont Jr., 186–203. Urbana: University of Illinois Press.

————. 1993. *Philosophy of Technology: An Introduction*. New York: Paragon House.

Kaltenborn, Olaf. 2001. "Der große Karneval: Im Cyberspace ist das ganze Jahr Fasching" [The great carnival: The whole year is Mardi Gras in cyberspace]. *Journal Phänomenologie* 15. Available online at <http://www.journal-phaenomenologie.ac.at/texte/jph15sp3.html>.

Kaufmann, Walter. 1974. Translator's introduction. In Friedrich Nietzsche, *The Gay Science: With a Prelude in Rhymes and an Appendix of Songs*, translated by Walter Kaufmann, 3–26. New York: Vintage Books.

Kirk, G. S., and J. E. Raven. 1966. *The Presocratic Philosophers: A Critical History with a Selection of Texts*. Cambridge: Cambridge University Press.

Landow, G. 1992. *Hypertext: The Convergence of Contemporary Critical Theory and Technology*. Baltimore: Johns Hopkins University Press.

————. ed. 1994. *Hyper/Text/Theory*. Baltimore: Johns Hopkins University Press.

Lyotard, J.-F. 1984. *The Postmodern Condition: A Report on Knowledge*, translated by G. Bennington and B. Massumi. Minneapolis: University of Minnesota Press. First published in 1979.

Maitland, Carleen, and Johannes Bauer. 2001. "National Level Culture and Global Diffusion: The Case of the Internet." In *Culture, Technology, Communication: Towards an Intercultural Global Village*, edited by Charles Ess, 87–120. Albany, N.Y.: State University of New York Press.

Mall, Ram Adhar. 2000. *Intercultural Philosophy*. Lanham, Md.: Rowman and Littlefield.

McLuhan, Marshall. 1965. *Understanding Media*. New York: McGraw-Hill.

Meyer-Drawe, Käte. 2001. "Im Netz" [(Caught) in the Net]. *Journal Phänomenologie* 15. Available online at <http://www.journal-phaenomenologie.ac.at/texte/jph15sp2.html>.

Miani, Mattia. 2000. "Civic Networks: A Comparative View." Internet Research 1.0: The State of the Interdiscipline. First Conference of the Association of Internet Researchers, University of Kansas, Lawrence, September 14.

Moravec, H. 1988. *Mind Children: The Future of Robot and Human Intelligence*. Cambridge, Mass.: Harvard University Press.

Negroponte, Nicholas. 1995. *Being Digital*. New York: Knopf.

Ong, W. J. 1981. *The Presence of the Word: Some Prolegomena for Cultural and Religious History*. Minneapolis: University of Minnesota Press.

———. 1988. *Orality and Literacy: The Technologizing of the Word*. London: Routledge.

Overholt, Thomas W., and J. Baird Callicot. 1982. *Clothed-in-Fur and Other Tales: An Introduction to an Ojibwa Worldview*. Lanham, Md.: University Press of America.

Postma, Louise. 2001. "A Theoretical Argumentation and Evaluation of South African Learners' Orientation towards and Perceptions of the Empowering Use of Information." *New Media and Society* 3, no. 3 (September): 315–28.

Sandbothe, Mike. 1999. "Media Temporalities of the Internet: Philosophies of Time and Media in Derrida and Rorty." *AI and Society* 13, no. 4: 421–34.

Shrader-Frechette, Kristen, and Laura Westra, eds. 1997. *Technology and Values*. New York: Rowman and Littlefield.

Sofield, T. 2000. "Outside the Net: Kiribati and the Knowledge Economy." In *Second International Conference on Cultural Attitudes towards Technology and Communication 2000*, edited by Fay Sudweeks and Charles Ess, 3–26. Murdoch, Australia: School of Information Technology, Murdoch University. Available online at <http://www.it.murdoch.edu.au/~sudweeks/catac00/>.

Sveningsson, Malin. 2001. "Creating a Sense of Community: Experiences from a Swedish Web Chat," Doctoral Dissertation, Tema Institute, Department of Communication Studies, Linköping, Sweden, 26–44.

Thanasankit, Theerasak, and Brian J. Corbitt. 2000. "Thai Culture and Communication of Decision Making Processes in Requirements Engineering." In *Second International Conference on Cultural Attitudes towards Technology and Communication 2000*, edited by Fay Sudweeks and Charles Ess, 217–42. Murdoch, Australia: School of Information Technology, Murdoch University.

Turk, Andrew, and Kathryn Trees. 1998. "Culture and Participation in Development of CMC: Indigenous Cultural Information System Case Study." In *Proceedings: Cultural Attitudes towards Communication and Technology '98*, edited by Charles Ess and Fay Sudweeks, 219–23. Sydney: Key Centre of Design Computing.

Sy, Peter. 2001. "Barangays of IT: Filipinizing Mediated Communication and Digital Power." *New Media and Society* 3, no. 3: 297–313.

U.S. Department of Commerce, Economics, and Statistics Administration, National Telecommunications and Information Administration. 2000. Falling through the Net: Toward Digital Inclusion. A Report on Americans' Access to Technology Tools. Available online at <http://www.ntia.doc.gov/ntiahome/fttn00/Falling.htm#2.1>. (Consulted September 8, 2001.)

Wheeler, Deborah. 2001. "New Technologies, Old Culture: A Look at Women, Gender, and the Internet in Kuwait." In *Culture, Technology, Communication: Towards an Intercultural Global Village*, edited by Charles Ess, 187–212. Albany, N.Y.: State University of New York Press.

Winograd, T., and F. Flores. 1987. *Understanding Computers and Cognition: A New Foundation for Design*. Reading, Mass.: Addison-Wesley.

Yoon, Sunny. 2001. "Internet Discourse and the Habitus of Korea's New Generation." In *Culture, Technology, Communication: Towards an Intercultural Global Village*, edited by Charles Ess, 241–60. Albany, N.Y.: State University of New York Press.

Zaharna, R. S. 1995. "Understanding Cultural Preferences of Arab Communication Patterns." *Public Relations Review* 21, no. 3 (Fall): 241–55.

15

HEURISTIC METHODS FOR COMPUTER ETHICS

WALTER MANER

Procedures are the stock-in-trade of computer professionals. Computer programs are most often written in procedural languages, and very often reduce to collections of procedures. The idea of an effective procedure, or algorithm, clearly remains the single most important concept in the field of computer science. Not surprisingly, then, many computer professionals are adept at procedural thinking. It has become their preferred cognitive style. This suggests that a step-by-step, or procedural, approach to ethical decision making could be an especially good fit for computer professionals.

I note, in passing, that this premise would likely be rejected by those, like John Rawls and Jeroen van den Hoven, who have argued that the aim of ethical analysis is not to complete a procedure. Rather, it is to reach a homeostatic cognitive state described by the phrase "wide reflective equilibrium." I differ from them in believing that this state can be incorporated into a procedural approach.

"Procedural ethics" may be a new concept, but it has a cousin in "procedural epistemology," a term apparently coined by John Pollock (Pollock 1998). He states that one main task of procedural epistemology is to describe procedures that, if they could be applied without constraint, would lead a rational intellect from input data to reasonable beliefs. By analogy, procedural *ethics*, as *its* first task, would seek to describe procedures that could play a similar role in guiding ethical reflection. In procedural epistemology, Pollock believes that once we have described all the useful procedures, we must next develop a "control structure" that will determine when each procedure can be used to best advantage. Procedural ethics would need a similar control structure; this would be its second task. This essay is mostly focused on the first task of procedural ethics but, at the end, I shall suggest how the second task could be approached.

"Algorithms" for Ethical Analysis

Ethical problems are too complex and too fluid to solve algorithmically in human time. If, however, one tried *per impossible* to construct an algorithm for ethical analysis, it might resemble this one:

```
Construct ethical-theory-list
Construct personal-virtue-list
Inspect situation
Construct shared-value-list
Construct list-of-parties

FOR each party in list-of-parties
| Construct option-list for this party
| FOR each option in option-list
| | Construct outcome-list for this option
| | FOR each outcome in outcome-list
| | | IF probability of this outcome is low
| | |   Delete this outcome from outcome-list
| | | ELSE IF moral relevance of this outcome is low
| | |   Delete this outcome from outcome-list
| | | ELSE
| | | | FOR each party in list-of-parties
| | | | | IF this party is affected by this outcome
| | | | |   Add this party to stakeholder-list

FOR each party in list-of-parties
| Recall option-list for this party
| FOR each option in option-list
| |
| | Recall shared-value-list
| | FOR each value in shared-value-list
| | | IF this option promotes this value
| | |   Increase weight of this option in option-list
| | | ELSE IF this option diminishes this value
| | |   Decrease weight of this option in option-list
| |
| | Recall personal-virtue-list
| | FOR each virtue in personal-virtue-list
| | | IF this option promotes this virtue
| | |   Increase weight of this option in option-list
| | | ELSE IF this option diminishes this virtue
| | |   Decrease weight of this option in option-list
| |
| | Recall outcome-list for this option
| | FOR each outcome in outcome-list
| | | IF benefits exceed risks for this outcome
| | |   Increase weight of this option in option-list
| | | ELSE IF risks exceed benefits for this outcome
```

```
| |  |  Decrease weight of this option in option-list
| |
| |  Recall stakeholder-list
| |  FOR each stakeholder in stakeholder-list
| |  | Construct list-of-obligations party has to stakeholder
| |  | FOR each obligation in list-of-obligations
| |  |  | IF option fulfills this obligation
| |  |  |   Increase weight of this option in option-list
| |  |  | ELSE IF option violates this obligation
| |  |  |   Decrease weight of this option in option-list
| |
| |  Recall ethical-theory-list
| |  FOR each theory in ethical-theory-list
| |  | Construct list-of-principles for this theory
| |  | FOR each principle in list-of-principles
| |  |  | IF option is consistent with this principle
| |  |  |   Increase weight of this option in option-list
| |  |  | ELSE IF option is inconsistent with this principle
| |  |  |   Decrease weight of this option in option-list
|
| Sort option-list by weight
| IF option-list[1] > option-list[2]
|   best-course-of-action = option-list[1]
| ELSE
|   best-course-of-action = Break-tie( option-list )
| Print party
| Print best-course-of-action
```

Does this pseudo-coded procedure supply at least a promising beginning for a complete, computable, effective procedure? No, not from a computational standpoint, not from an ethical standpoint, and not even from a structural standpoint.

A computer scientist would notice the omission of important programmatic elements from this "program":

- It makes no provision for asserting or withdrawing assumptions.
- It makes no provision for asserting or withdrawing facts.
- It makes no provision for asserting or withdrawing hypotheses ("what-if" analysis).
- It makes no provision for backtracking from dead ends.
- It makes no provision for calculating the values used to adjust the weights in the option list.
- It makes no provision for exiting early if an overriding or pivotal consideration emerges.

```
|  |  | ELSE IF this option diminishes this value
|  |  |   Decrease weight of this option in option-list
|  |
|  | Recall personal-virtue-list
|  | FOR each virtue in personal-virtue-list
|  |  | IF this option promotes this virtue
|  |  |   Increase weight of this option in option-list
|  |  | ELSE IF this option diminishes this virtue
|  |  |   Decrease weight of this option in option-list
|  |
|  | Recall outcome-list for this option
|  | FOR each outcome in outcome-list
|  |  | IF benefits exceed risks for this outcome
|  |  |   Increase weight of this option in option-list
|  |  | ELSE IF risks exceed benefits for this outcome
|  |  |   Decrease weight of this option in option-list
|  |
|  | Recall stakeholder-list
|  | FOR each stakeholder in stakeholder-list
|  |  | Construct list-of-obligations party has to stakeholder
|  |  | FOR each obligation in list-of-obligations
|  |  |  | IF option fulfills this obligation
|  |  |  |   Increase weight of this option in option-list
|  |  |  | ELSE IF option violates this obligation
|  |  |  |   Decrease weight of this option in option-list
|  |
|  | Recall ethical-theory-list
|  | FOR each theory in ethical-theory-list
|  |  | Construct list-of-principles for this theory
|  |  | FOR each principle in list-of-principles
|  |  |  | IF option is consistent with this principle
|  |  |  |   Increase weight of this option in option-list
|  |  |  | ELSE IF option is inconsistent with this principle
|  |  |  |   Decrease weight of this option in option-list
|
| Sort option-list by weight
| IF option-list[1] > option-list[2]
|   best-course-of-action = option-list[1]
| ELSE
|   best-course-of-action = Break-tie( option-list )
| Print party
| Print best-course-of-action
```

- It makes no provision for exiting early when further analysis would be a waste of time.
- It makes no provision for simplification except in the case of deleting outcomes.

A philosopher would notice that the "algorithm" omits important techniques commonly used in ethical analysis:

- It makes no provision for reasoning by example, including the posing and evaluation of counterexamples.
- It makes no provision for case-based reasoning, omitting even a simple appeal to precedent.
- It makes no provision for conflict resolution except naively, in the case of numerical ties.
- It makes no provision for moral intuition.
- It makes no provision for semantic or conceptual analysis.
- It makes no provision for the elaboration or testing of moral analogies.
- It makes no provision for the elaboration or testing of moral arguments.

Beyond this, there are severe structural problems:

- The steps should be logically independent but clearly are not. An action taken by one party may introduce or eliminate stakeholders. Similarly, an action taken by one party may limit or expand the options available to another party.
- The procedure makes the dubious assumption that moral evaluations are scalable and, in fact, that evaluations of diverse origins can be directly compared on a common scale.

Despite this abundance of glosses and omissions, this "algorithm" still seems to *overspecify* the process of ethical analysis, at least from a human standpoint. If there were only five items to consider in each major list, then there would be 5^4 subcases to process at three points, and 5^3 subcases to process at four points, for a grand total of 2,375 subcases. Maybe a computer could help; I shall return to that thought later.

Heuristics for Ethical Analysis

Technically, an algorithm is an *effective* procedure, which means that it will unfailingly produce the intended result in a finite number of steps *if* it receives sufficient valid input. Algorithms can be contrasted with *heuristics*, which are stepwise procedures that will *tend* to produce the intended result when they get the right input. If I have misplaced my keys, a useful heuristic would be to examine my immediate environment on the assumption that the item is more likely to be nearby. An algorithm, by contrast, would require an examination of all environments that I have encountered since I first acquired the keys.

I believe that the procedures proposed for guiding ethical analysis can be valuable heuristically even when they are inadequate as algorithms.

The search for useful analytical heuristics has been a common theme in applied ethics for many years. More than sixty procedures are detailed on the web pages created to support this essay: http://csweb.cs.bgsu. edu/maner/heuristics. Within computer ethics, heuristics have been of early and continuous interest.

Although human beings make moral judgments from a very young age, heuristics may be needed

- to take this process to a higher level,
- to guard it against omissions,
- to make it more teachable, or
- to ensure that an effective process, once found, can be repeated and refined.

Accordingly, heuristics have been invented to direct case studies, to focus ethical scenarios, and to frame solutions for moral dilemmas. Professional organizations have published heuristics to raise professional awareness and to transmit professional culture. Ethical codes sometimes include heuristics as a proof that the code *does* translate into practice. At a higher level, heuristics have emerged as a by-product of attempts to apply, or even to validate, specific ethical theories. In short, heuristic methods supply part of the core meaning for "applied" ethics.

Organization of This Essay

From the five dozen procedures I have collected and studied, I present twelve that exhibit a variety of approaches. Then I show how individual steps, collected from all these many procedures, can be mapped into a uniform set of stages. Next I show how steps that map into a *particular* stage could be reformulated more usefully as "checkpoints." I then identify faults common to many of the procedures I have collected. Next I suggest how my stage-by-stage decision-making model could be re-purposed, adapted, filtered, or downsized for use in specific circumstances. Finally, I explain what makes the procedural approach to ethical decision making different when used in the field of computer ethics.

A Sampling of Representative Procedures

Note. Most of these procedures contain both steps and substeps, but all the substeps have been omitted here to achieve a more compact presentation. Complete versions can be found at the web address given earlier.

The Practicum Method (Maner 1981)

1. Form an "ethics committee" of at least five persons.
2. Frame a specific question that creates the desired ethical dilemma.
3. Construct a scenario (ethical story or vignette) of about 150 words that will evoke the dilemma.
4. Construct at least three persuasive arguments on each side of the question (yes and no).
5. Raise objections to these arguments.
6. Make replies to the objections.
7. Make counterreplies to these replies.
8. Take a stand on the issue. Reach a "verdict of one."

This procedure requires seventy-five steps in its complete form, but the condensed version shown above is sufficient to exhibit the distinctive elements of the approach, which include

- the creation of an issue as a *first* step, followed by the construction of a matching scenario,
- a strong "committee" or collaborative emphasis in all but the last step, and
- an attempt to explore the issue through the vehicle of structured debate.

The main constraint on the use of this method is the requirement that participants be experienced in philosophical argumentation and debate.

A Strategy for Solving Moral Problems (McLaren 1989)

1. Formulate the problem.
2. Propose a hypothetical solution.
3. Explain what means and consequences will be involved in accepting the hypothesis.
4. State every important reason for accepting the hypothesis and those for rejecting it.
5. Decide by weighing the importance and certainty of the reasons.

At the top level, this method is generic enough to be useful in almost any problem-solving activity, including problems outside ethics. I conclude from this that McLaren views ethical decision making as a specialization, or refinement, of the general process of rational decision making. While his method includes some consequentialist elements, it mostly follows the "good reasons" approach to morality advocated by Stephen Toulmin, Kurt Baier, and, more recently, James Rachels. Users of this method must be skilled in the discovery, elaboration, and comparative evaluation of moral reasons.

Canadian Psychological Association (1991)

1. Identification of ethically relevant issues and practices.
2. Development of alternative courses of action.
3. Analysis of likely short-term, ongoing, and long-term risks and benefits of each course of action on the individual(s) or group(s) involved or likely to be affected.
4. Choice of course of action after conscientious application of existing principles, values, and standards.
5. Action, with a commitment to assume responsibility for the consequences of the action.
6. Evaluation of the results of the course of action.
7. Assumption of responsibility for consequences of action, including correction of negative consequences, if any, or re-engaging in the decision-making process if the ethical issue is not resolved.

Most codes of ethics include sections on principles and professional practices. The CPA Code of Ethics includes these predictable elements, but it also includes a decision-making model designed to put these principles and practices into action. The recommended procedure is remarkable for giving nearly equal weight to steps required before and *after* a decision has been rendered. The *post*-decision process includes

- the explicit assumption of personal responsibility for the decision and its consequences,
- the monitoring of the results with an eye toward mitigation of negative consequences, and
- an unusual requirement to re-engage the decision-making process if the ethical issue is not resolved.

Those who adopt this approach cannot dismiss past decisions, even carefully rendered ones, as water under the bridge; with this method, the bridge moves with the water.

The Paramedic Method (Collins and Miller 1995)

1. Gather data systematically about the parties.
2. Analyze the data systematically for the alternatives.
3. Try to negotiate a social-contract agreement in an imaginary meeting where all parties are represented.
4. Judge each of the alternatives according to ethical theories.

For steps 1 and 2, "systematically" may be an understatement. Although not apparent in this condensed version, the Paramedic Method requires *meticulous* attention to detail. Like the "algorithm" for ethical analysis that I constructed earlier, this procedure iterates computer-like over

- each involved party,
- each pair of involved parties,
- each linkage of rights with corresponding duties for each pair of involved parties,
- each course of action open to each party,
- each risk and benefit for each action for each party, and
- each ethical theory that may bear on each action of each party.

Together these tightly knit iterations seem to impose a discipline that demands, of the user, a degree of precision and thoroughness usually reserved for the construction of mathematical proofs. However, in the give and take of the imaginary negotiation done in step 3, the problem may reduce to a smaller set of critical considerations. If not, users will need a bookkeeping system to track all the decision elements that come into active play.

Nine Checkpoints for Ethical Decision Making (Kidder 1995)

1. Recognize that there is a moral issue.
2. Whose issue is it?
3. Gather the relevant facts.
4. Test for right-versus-wrong issues.
5. Test for right-versus-right paradigms. What sort of dilemma is this?
6. Apply the resolution principles.
7. Investigate the "trilemma" options.
8. Make the decision.
9. Revisit and reflect on the decision.

Rushworth Kidder's approach is notable for distinguishing between unconflicted decisions ("right versus wrong") and the more problematic, *conflicted* decisions ("right versus right"). The resolution of right-versus-wrong conflicts is trivial, but right-versus-right conflicts are inherently difficult because they can pit

- justice against mercy,
- freedom against security,
- short-term consequences against long-term,
- the individual against the community, or
- truth against loyalty.

To resolve this second, more difficult type of conflict, Kidder applies ethical principles and, where that does not produce a complete resolution, he tries to find a "creative middle ground" that will convert the dilemma into a "trilemma." Kidder's book is called *How Good People Make Tough Choices*, and it is highly recommended.

Worksheet for Ethical Decision Making (Bivins 1996)

---

1. What is the ethical issue or problem?
2. What immediate facts have the most bearing on the ethical decision you must render in this case?
3. Who are the claimants in this issue, and in what way are you obligated to each of them?
4. What do you think each of these claimants would prefer that you do regarding this issue?
5. List at least three alternative courses of action.
6. Are any of your alternatives supported or rejected by ethical guidelines?
7. Determine a course of action based on your analysis.
8. Defend your decision in the form of a letter addressed to your most adamant detractor.

This procedure includes an "in their shoes," or empathy, step that requires the decision maker to examine possible outcomes through the eyes of each stakeholder. This clearly requires creativity, imagination, a fair grasp of human nature, and the ability to project oneself into the lives of other people. The same qualities are needed again in the last step, where the decision maker tries to address the concerns of an imaginary detractor.

Risk/Benefit Model (Hiskes 1996)

---

1. Identify the problem and basic policy objectives.
2. Formulate alternative courses of action.
3. Identify relative consequences of each alternative.
4. Assign a probability to each relevant consequence.
5. Assign a value, i.e., a numerical cost or benefit, to each consequence.
6. Combine the information obtained in steps 3 to 5 and select the best alternative.

Various decision-making procedures require rankings, which is an elementary form of enumeration, but none of the methods I examined is as bold with numbers as Hiskes is. With this particular approach, likelihoods are quantified along with costs and benefits, and then all these numbers are "combined" using some unspecified formula. Any person using Hiskes's method would have to be comfortable taking a thoroughly quantitative approach. Numerics have the best chance to work within a consequentialist framework, which Hiskes clearly favors.

Ethical Decision-Making Model (Waldfogel 1996)

---

1. Identify stakeholders.
2. Identify values: ethical and nonethical.
3. Ethical values trump nonethical values.
4. If two ethical values conflict, the one that produces the greatest good for the greatest number wins.

These four steps are a *complete* listing of the procedure recommended by the Josephson Institute for use by school-age youth in their teen years. Although clearly minimalist in its content, the procedure still manages to include *two* provisions for conflict resolution. It is difficult to imagine how a defensible ethical decision could be made using a procedure any simpler than this one. For the intended audience, the procedure should work, provided its users have been taught to distinguish between ethical and nonethical values.

The Five-Step Process of Ethical Analysis (Rahanu, Davies, and Rogerson 1996)

---

1. Analyze the case.
2. Apply formal guidelines.
3. Apply ethical theories.
4. Apply relevant laws.
5. Apply informal guidelines.

It is interesting that informal guidelines are used at the very *end* of this procedure, after the hard work is complete and a tentative course of action has been determined. As the critical moment approaches, just in case the high-powered machinery of ethical analysis may somehow have blundered, a last-ditch battery of common-sense tests are applied:

- The *Mother Test*:
  Would you tell her? Would she be proud or ashamed?
- The *TV Test*:
  Would you tell a nationwide audience of your actions?
- The *Smell Test*:
  Do you feel "in your bones" that there is a problem?
- The *Other Person's Shoes Test*:
  What if the roles were reversed?
- The *Market Test*:
  Could you advertise the act to give yourself a marketing edge?

If the contemplated action is confirmed by these last tests, it may be a sign that the decision maker has reached the desirable state of "wide reflective equilibrium" important to some moral philosophers.

Ethical Decision Making (Gregoire 1997)

---

1. What are the facts?
2. What ethical principles should be applied?
3. Who should decide?
4. Who should benefit from the decision?
5. How should the decision be made (implemented)?
6. What steps should be taken to prevent this issue from occurring again?

This procedure pushes to the very top level questions that are mostly ignored by other approaches. Instead of applying a fixed set of ethical principles to every situation, this procedure makes a context-dependent selection. This could be a first step toward the "control structure" I mentioned earlier. Gregoire's procedure is also interesting because, instead of ignoring implementation details, it considers implementation as a moral issue in its own right and aims to achieve due process for all involved parties. Finally, there is a healthy emphasis on prevention not found in other ethical decision-making models.

Quick Tests for Ethical Congruence (Ethics Resource Center 1998)

1. How does your stomach feel?
2. Are you bothered or upset by the decision you are about to make?
3. Do you have doubts?
4. Do you wish you didn't have to choose?
5. Is it frustrating to have to select one option over the others?
6. How would you feel if your mother were looking over your shoulder?
7. How would you advise your own child to act in this same set of circumstances?
8. Could you accept public review of the tradeoffs and compromises you made?
9. Could you accept review by friends, neighbors, and family?

In my view, these tests address two very different circumstances, one that comes after a long period of analysis and another that comes when no analysis is possible. When a person comes to the end of a difficult and conflicted process of ethical deliberation, when the only remaining step is to put the decision into action, what can the decision maker do to confirm the choice?

Questions 1 through 5 may provide an answer. On the other hand, when an immediate decision is required, when there is no lead time for normal ethical deliberation, what can the decision maker do to reduce the risk of error? Questions 6 through 9 may provide an answer.

The HARPS Methodology (Searing 1998)

1. Collect information.
2. Select a moral agent.
3. List known relevant facts.
4. Make factual assumptions.
5. List conceptual issues.
6. Define these concepts.
7. List moral issues.
8. Select one or more methods of analysis.
9. Negotiate a conclusion based on the solutions provided by the methods of analysis.

The fifty-three-step HARPS Methodology is one of the most comprehensive approaches to ethical decision making ever devised. It is so thorough that it comes with its own fifty-page user manual. In fact, it is probably too complete a method for reliable use by ordinary human beings. Not to worry. HARPS has been incorporated into the Ethos System, a computer program that guides the user through the labyrinth one step at a time:

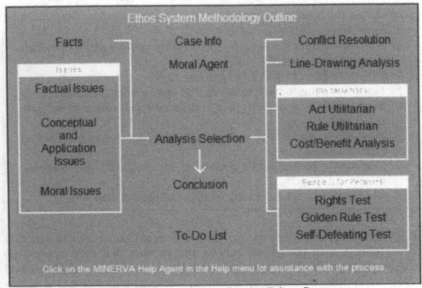

Main Navigation Screen for the Ethos System

HARPS and Ethos together offer a good fit in the situation where the decision maker has time for extended deliberation, particularly where stakes are high and the deliberation must be conducted with maximum thoroughness. It also offers a good fit where decisions must be accompanied by a complete written justification, one that shows in detail exactly how and why the decision was made. With the Ethos System, this is as simple as choosing Print from the program menu.

The HARPS Methodology emerged from the Engineering Ethics program at Texas A and M University, and it is based on the work of Charles E. Harris, Michael S. Pritchard, and Michael J. Rabins. I highly recommend their book, *Engineering Ethics: Concepts and Cases*.

A Uniform Set of Stages

I concede that the best minds in applied ethics will never agree on a uniform set of steps and substeps for their heuristic decision-making procedures. We all have our creative differences, we all build our procedures for different audiences, and we all use our methods to address the different problems and issues. At a higher level of abstraction, however, it may be possible to reach consensus on a uniform set of *stages*. As a step in that direction, I propose the following:

1. *The Preparing Stage.* During this earliest stage, we cultivate moral awareness and sensitivity, clarify our value system and worldview, observe human nature, engage in ethical behavior ourselves, learn some ethical theory, and prepare to avoid ethical traps.

2. *The Inspecting Stage.* We now face a possible problem situation, so we attempt to define the problem by noting facts, participants, groups, roles, relationships, events, and actions. At this stage, we make no effort to determine what is *morally* relevant, only what is *factually* relevant, and we try to produce a list of uncontroverted facts that would be acceptable to all parties. Finally, we determine whether the situation is truly a problem, one that requires further attention and action.

3. *The Elucidating Stage.* Here we identify facts that are missing, and either develop these new facts or make assumptions to cover them. We clarify technical, ambiguous, or vague concepts, and we try to eliminate biased and emotionally charged language. We isolate key factors in the situation, including especially factors that set this situation apart from otherwise similar cases. We identify the difficulties and obstacles that may hinder analysis. We do epistemological legwork, trying to assess the reliability of our sources and the validity of our information. We try to determine the immediate antecedents of the problem, how the situation came to be. We discriminate between primary and secondary participants, and we determine which parties are affected by actions of other parties. Potentially, these affected parties are the stakeholders, but we do not make that association until the Focusing Stage. We consider all the lists we have made and eliminate from these lists any items that do not meet some minimum threshold for significance. Given all this, we try to estimate whether this is a short-term problem that can be resolved quickly, or a long-term problem that requires sustained effort. Finally, in a very preliminary way, we begin to frame possible issues: "Is it true that X should do Y assuming Z?"

4. *The Ascribing Stage.* We begin to infer and specify the values, goals, ideals, interests, ideologies, priorities, and motives that are most likely responsible for creating the dynamics of the problem. We ascribe these biases, tendencies, and proclivities to various participants or to ourselves.

5. *The Optioning Stage.* We brainstorm to list all possible courses of action that are (or were) available to the participants. This list may include actions that are ill-advised and actions that are contingent on other actions. Once we know the full range of alternatives, we try to eliminate from the list those actions that are clearly not feasible or that

fail to meet some threshold for relevance. We do not exclude an option because we think it is wrong.

6. *The Predicting Stage.* For each remaining option, we list potential consequences, including consequences that would result if no action were taken. We discriminate between short- and long-term effects, between likely and unlikely consequences, and between results that are intended and unintended. We associate these consequences with specific participants or with ourselves, either as a risk or as a benefit.

7. *The Focusing Stage.* We consider all affected parties and identify those who are sufficiently affected to be elevated to stakeholder status. We note the rights that are claimed, or could be claimed, and we identify the responsibilities or duties that correspond to those rights. We determine which facts are morally relevant, which actions have moral consequences, which values are moral values, which questions are moral questions, and which issues are moral issues. We take special note of virtues, values, rights, priorities, and ideals that appear to be at risk or that appear to be in conflict. We eliminate all factors that are morally irrelevant or insufficiently relevant. Based on all this analysis, we identify and define the core ethical issue, which is often expressed as a dilemma: "Should X do or not do Y assuming Z?"

8. *The Calculating Stage.* Some decision-making procedures attempt to quantify risks, costs, benefits, burdens, impact, likelihoods, and even relevance. These weights and numbers, if required, are generated at this stage. Later, at least in theory, it will be easy to determine which option produces the most probable morally relevant benefit, with the least probable morally relevant risk.

9. *The Applying Stage.* This is the stage where most of the critical work of applied ethics is done. Ideally, each possible stakeholder/action pair is considered separately and sympathetically. Reasons for and against particular actions are cataloged, then ranked. Morally required actions are distinguished from those that are morally permitted but not required. Values are weighed against other values. Sometimes entire value systems are weighed against competing systems. Short-term benefits are weighed against long-term risks. In similar fashion, long-term benefits are weighed against short-term risks. Various ethical theories may come into full play—and into full conflict. Like and unlike cases are considered and compared. We construct moral analogies and dis-analogies, examples and counterexamples. Best- and worst-case scenarios are elaborated. Diverse ethical principles are applied, and we note whether their advice is conflicting or convergent. Options are evaluated according to the virtues they promote, or the rights they respect, or the obligations they satisfy, or the values they maximize, or the principles they obey. Philosophical arguments are constructed, deconstructed, and evaluated. Laws, policies, ethical codes, and professional literature are reviewed for parallels. Associates,

supervisors, mentors, trusted friends, advisers, and stakeholders (if willing and available) give the decision maker the benefit of their opinions.

Results may be convergent but typically are conflicting, contradictory, or inconsistent. Because conflicts are so common, special strategies are invoked to resolve them. When the dust settles, we hope the problem has been reduced to a coherent set of pivotal considerations. If this happens, the long list of options produced in stage 5 can be shrunk to a much shorter list of promising options. For these remaining options, full justifications are prepared.

10. *The Selecting Stage.* An option is chosen, and that choice is confirmed by applying a series of informal, common-sense ethical tests (e.g., the Reversed Roles Test or the Public Scrutiny Test). As a double check, we may perform a "sensitivity analysis" to identify those situational factors that, if altered, would greatly alter our decision. We would then revisit our analysis of those factors. All things considered, we may not be 100 percent comfortable with our decision. Even so, we should reach a settled state of "wide reflective equilibrium." If not, we may decide to restart the analysis at an earlier stage, time permitting.

11. *The Acting Stage.* We plan exactly what is to be done step by step, and who is to do it. We try to ensure due process for all stakeholders. We may construct a time line to sequence individual actions. We identify the means to be employed. We gather the necessary resources. We develop indicators of success and failure, including some early indicators. Finally we take action, and we take responsibility for the consequences.

12. *The Reflecting Stage.* In this final stage, we monitor the decision as it is implemented with special attention to the effects it is having on stakeholders. We assess the results as they unfold, using the indicators developed in the previous stage. If, by those early indicators, the decision is failing, we may re-implement the decision, and if that also fails, we may abort and start over, circumstances permitting. Otherwise we live with the decision and learn from it. When finished monitoring, we may recommend new policies to address particular issues. We may review and evaluate the decision procedure itself with an eye toward process improvement. Did the procedure work as intended? Were the steps in the correct order? Finally, we may consider what could have been done in the first place to prevent the problem and, where appropriate, take steps to prevent recurrence.

Time and other resources permitting, a complete decision-making procedure should conduct us through at least eleven of these twelve stages, possibly omitting the Calculating Stage. Do the available procedures provide good coverage of these stages? Apparently not. The sixty-plus decision-making procedures I studied give very uneven emphasis:

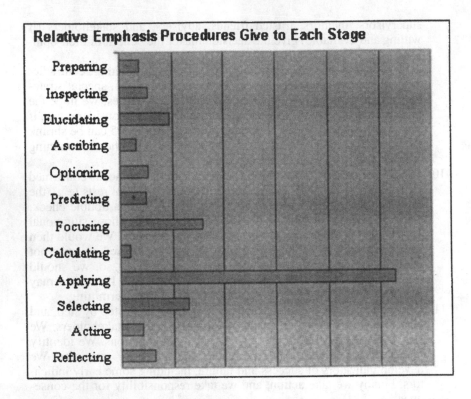

Relative Emphasis Procedures Give to Each Stage

- Preparing
- Inspecting
- Elucidating
- Ascribing
- Optioning
- Predicting
- Focusing
- Calculating
- Applying
- Selecting
- Acting
- Reflecting

Most concentrate on the Applying Stage, where the critical work of applied ethics is done. Nearly all omit one or two stages, and half omit three or more.

Reformulating Stages as Checkpoints

Can we consolidate all of the available procedures into one omnibus all-purpose ethical decision-making procedure? Probably not, but it might be possible to convert them into sets of "checkpoints," one set for each of the twelve stages. Unlike the serialized steps included in true procedures, checkpoints are compromised by the fact that they are weakly ordered. To illustrate what may be possible, I offer a consolidated set of checkpoints for stage 1:

Consolidated Checkpoints for the Preparing Stage

Advance Preparations

1. Become aware of beliefs that can create ethical traps, including

- the belief you can reach objectivity without consulting others,
- the belief that there are no right or wrong answers, and
- the belief that personal values can trump ethical principles or professional standards.

2. Build the ego strength necessary to maintain your personal integrity in the face of external pressures.
3. Cultivate such personal virtues as integrity, honesty, fidelity, charity, responsibility, and self-discipline.
4. Define your personal worldview, including your concept of an ideal and just society.
5. Develop a circle of advisers.
6. Engage in ethical behavior and profit from your mistakes.
7. Learn some ethical theory.
8. Learn to recognize the moral dimension present in *ordinary* human experience. (Almost everyone sees the big issues.)
9. Listen to your conscience or spiritual center.
10. Observe the causes and consequences of human behavior.
11. Practice being the devil's advocate or engage in other forms of ethical role-playing.
12. Review ethical codes.
13. Take advantage of opportunities to educate and elevate your moral sensitivities.
14. Take inventory of your personal values, with special attention to ethical values.
15. Test your ethical ruminations by exposing them to trusted friends and mentors.

Last-Minute Preparations

1. Decide how best to channel your immediate, intuitive, prereflective responses.
2. Estimate how much time can safely be spent in analysis before the decision comes due.
3. If very little time is available, consider restructuring the situation to allow more time for deliberation.
4. Formulate a goal that captures the desired outcome for the entire decision-making process.

Faults Common to Many Ethical Procedures

Most of the procedures I studied were designed to meet specific needs for specific persons in specific kinds of situations. They were not offered as exhaustive or universal formulas. Still, for the sake of process improvement, it is useful to consider their limitations.

These appear to be the most serious limitations:

- Many cannot deal with situations that change rapidly while they are under analysis.
- Many define the ethical issue too early in the process.
- Many do not degrade gracefully under pressure of time.
- Many do not recognize destructive interactions between steps (e.g., when later steps invalidate earlier steps).
- Many expand to several thousands of small steps when used in complex situations.
- Many implement only one approach to ethics (e.g., utilitarianism).
- Many make no provision for regression (backtracking) or for redoing a step.
- Many need elimination, simplification, or other types of problem-reduction steps.
- Many require a high level of situational ethical awareness in the very first step.
- Many try to determine moral relevance too early in the process.

The remaining limitations are less serious:

- Many cause conflicts and ambiguities to surface but offer no effective way to deal with these problems.
- Many could benefit from doing certain steps in parallel.
- Many do not allow a fact or an assumption to be withdrawn once it has been introduced.
- Many expect (or force) the issue to break into a dilemma, with exactly two basic alternatives.
- Many fail to offer a quick resolution once a pivotal consideration has emerged.
- Many have steps that appear to be out of order, in violation of apparent prerequisites.
- Many provide no way, in later stages, to update information developed at earlier stages.
- Many require special training or knowledge.
- Many weigh or rank considerations without telling us how the weighing or ranking is to be done.
- Many were designed for academic or training settings, and would not be appropriate for use in the field.

Adapting the Stage-by-Stage Model

Imagine that we have complete sets of checkpoints for all twelve stages, possibly containing several hundred individual checkpoints. Should every person in every situation consider every checkpoint? Clearly not. Situations differ, time pressures vary, and decision makers bring widely different backgrounds to the task. Pollack is right. There must be a "control strategy" that repurposes, adapts, filters, and downsizes these checkpoints for use in specific circumstances.

Presumably, to make this possible, we would follow three steps. First, we would specify exactly which checkpoints are usable under which constraints. For example, a particular checkpoint might not be usable by someone with no knowledge of ethical theory. Second, we would identify the specific constraints under which the user must operate. Finally, we would apply the user's constraints to the full sets of checkpoints, filtering out elements that violate those constraints. If this user had no knowledge of ethical theory, then any checkpoint requiring such knowledge would be eliminated.

In computer-science terms, this would require constructing a constraint-propagation network, one where each node is an individual checkpoint and clusters of nodes represent decision-making stages. Feasible, but very complicated. Gathering constraints would be simpler. We could use a web page to survey users:

I need to make a decision (check one)
( ) right now
( ) within a few hours
( ) today
( ) within a few days

I do not want to consider (check all that apply)
[ ] advice from others
[ ] corporate codes of conduct
[ ] corporate policies
[ ] ethical theory X
[ ] ethical theory Y
[ ] ethical theory Z
[ ] formal ethical principles (e.g., universalizability)
[ ] ideals and goals
[ ] informal tests of ethical convergence (e.g., the Mom Test)
[ ] laws and legal precedents
[ ] long-term consequences
[ ] moral analogies (examples and counterexamples)
[ ] motives and intentions
[ ] professional codes of conduct
[ ] rights and responsibilities
[ ] short-term consequences
[ ] values and value systems
[ ] virtues

I want to <u>avoid</u> (check all that apply)
[ ] applying general ethical principles
[ ] constructing philosophical arguments
[ ] debating the issues (making objections and replies)
[ ] doing semantic or conceptual analysis
[ ] predicting consequences
[ ] ranking consequences
[ ] ranking rights or responsibilities
[ ] ranking values or value systems
[ ] ranking virtues
[ ] steps that generate confusion, ambiguity, or conflict
[ ] surmising intent
[ ] using case-based, precedent-based, or example-based reasoning
[ ] using numeric and quasi-numeric methods

In the present situation, I have <u>no</u> access to (check all that apply)
[ ] advisers, consultants, mentors, or role-players
[ ] fact-finding or investigative resources
[ ] participants or stakeholders

I would describe this situation as one that requires (check all that apply)
[ ] a comprehensive defense or formal justification
[ ] a decision to be implemented (not just a training exercise)
[ ] a model decision that can serve as a precedent for future decisions
[ ] a procedure that can adjust to circumstances as they change
[ ] consensus or participatory decision making
[ ] extreme care (very high stakes)

Properly programmed, a computer could use this information to custom-build a procedure that both meets the needs of the user and fits the characteristics of the situation. This made-to-order procedure may not be structured for optimum results, especially if multiply constrained, but it could still produce situationally useful results. *For a heuristic procedure, this is enough.*

Procedural Computer Ethics

I have argued elsewhere (Maner 1996) that the field of computer ethics is driven by unique issues that derive from properties that make computers unique. Does this mean that *procedural* computer ethics will be unique as well?

There are some unique elements, but they come into play at the middle and lower levels of analysis. At the top level, decision-making procedures useful in computer ethics are not much different from procedures useful in other areas of applied ethics. And these, in turn, are not much different from generic procedures that could be used in areas outside ethics. So what are the unique decisional factors in computer ethics?

Computers may exacerbate a decision-making problem. The infusion of computer technology often makes the problem worse. Consider, for example, issues of copyright infringement. It is clearly wrong for me to copy a protected image to my web site without getting permission from the copyright holder. Suppose, however, I do not copy it but merely *link* to it, so that when my web page is rendered, the protected image pops into place within "my space." This benefits me, not the copyright holder. Am I morally entitled to this benefit?

Computers may multiply the number and location of consequences. Because computer processors execute their instructions so fast, they can produce many thousands of local consequences in the space of a few seconds. With only a little more time, computers can exploit network connections to reproduce these consequences at many thousands of remote locations. Computer viruses wreak havoc locally because computers are fast, then wreak havoc globally because computers are connected.

Computers may multiply the number of affected parties. If computers multiply consequences, then it is likely that these consequences will affect more people. Consider, for example, those database-management systems that hold our personal records. One unauthorized query run against a large database can affect millions of persons.

Computers may be required to implement a decision. Computer-driven problems may need computer-driven solutions. If, in a particular situation, we decide that we are morally obligated to provide better privacy protection for medical records, then computers may be required to provide the necessary encryption, authentication, and access control.

Computers may make some consequences more difficult to predict. Suppose, for example, that we willingly surrender several pieces of personal information for two very different business purposes. The data collected for these different uses are mostly different. In each case, we carefully avoid disclosing enough information to expose our identity. So far, so good. But if these data, separately harmless, find their way into databases that share information, then the *merged* information may lead direct to us. For most people, this consequence is difficult to envision, let alone to predict. In general, the behavior of computers is difficult to predict because there can be a disproportionate and discontinuous connection between cause and effect. A satellite may fail to deploy because, buried in a half-million lines of code, one instruction contains one bad character. Failures can be extremely generalized for problems that are extremely localized. This is a special problem for consequentialists, as they must be able to predict short- and long-term effects.

Computer-related concepts may be more technical or more difficult to elucidate. I observe with some sadness that my profession seems adrift in a shifting sea of technical jargon. We mostly have ourselves to blame, but in all fairness I must say that some concepts are inherently obscure. Consider, for example, the concept of a parity bit or an IP address or a digital watermark or a public key or a mutable virus. These concepts are not difficult because of jargonizing; they are difficult because they were born difficult.

Finally, computers may increase the attention given to certain types of rights. What rights receive extra attention? In an information-centric discipline like computer ethics, they are

- the right to access information,
- the right to block access to information (privacy), and
- the right to own and manage information (intellectual property).

For medical ethics, *life* is the primary good. For computer ethics, the primary good is *information*.

References

Agapow, Paul-Michael. 1998. "Analysis of Scenarios." http://www.cs.latrobe.edu.au/~agapow/Teaching/Cs292/pd.week.6.html (10 June 1999).

American School Counselor Association. 1996. "Ethical Decision Making Model." http://www.schoolcounselor.org/Ethics/ethics_d.html (9 June 1999).

Baase, Sara. 1997. "Issues of Professional Ethics and Responsibilities." In *A Gift of Fire: Social, Legal, and Ethical Issues in Computing*, 331–37. Upper Saddle River: Prentice-Hall.

Bivins, Tom. 1996. "Worksheet for Ethical Decision Making." http://jcomm.uoregon.edu/~tbivins/J397/Worksheet.html (6 Jan. 2002).

Black, Jay, Bob Steele, and Ralph Barney. 1999. "Making Ethical Decisions." In *Doing Ethics in Journalism: A Handbook with Case Studies*, 51–63. 3d edition. New York: Allyn and Bacon.

Campbell, C., S. Donnelly, B. Jennings, and K. Nolan. 1990. *New Choices, New Responsibilities: Ethical Issues in the Life Sciences*. Briarcliff Manor: Hastings Center.

Canadian Psychological Association. 1991. "The Ethical Decision-Making Process." In *Canadian Code of Ethics for Psychologists*. http://www.cpa.ca/ethics.html#ethic (7 June 1999).

Chonko, Lawrence B. 1995. "Decision Making When Ethical

Considerations Are Involved." In *Ethical Decision Making in Marketing*, 62–92. Thousand Oaks: Sage.

Collins, W. Robert, and Keith W. Miller. 1995. "Paramedic Ethics for Computer Professionals." In *Computer Ethics and Social Values*, edited by Deborah G. Johnson and Helen Nissenbaum, 39–56. Englewood Cliffs: Prentice-Hall.

Darty, Mark. 1998. "The Ethical Decision-Making Process." http://grants.cohpa.ucf.edu/age-ethics/ethical_process/ (10 June 1999).

Dreilinger, Craig, and Dan Rice. 1995. "Ethical Decision Making in Business." In *Business Ethics: Readings and Cases in Corporate Morality*, edited by W. Michael Hoffman and Robert E. Frederick, 89–93. 3d edition. New York: McGraw-Hill.

Dunlap, Joanna C. 1998. "Ethical Decision-making in Business." http://www.cudenver.edu/~jdunlap/5990/project1obj.html#sequence (8 June 1999).

Engler, Arthur J. 1998. "Summary of Ten-step Decision Model." http://www.nursing.ab.umd.edu/mch/courses/418/nuintro.htm (8 June 1999).

Ethics Resource Center. 1998. "The Big PLUS in Ethical Decision Making." http://www.ethics.org/abigplus.html (6 Jan. 2002).

———. 1998a. "Three Quick Tests for Ethical Congruence." http://www.lmco.com/erc/1-91b.html (9 June 1999).

Fitzpatrick, Dan. 1998. "An Ethical Decision Making Model." 1998. http://www.skidmore.edu/~dfitzpat/death/ethics.txt (10 June 1999).

Forester-Miller, Holly, and Thomas Davis. 1996. "A Practitioner's Guide to Ethical Decision Making." http://www.counseling.org/resources/pracguide.htm (8 June 1999).

Gregoire, Tom. 1997. "Ethical Decision Making." http://www.csw.ohio-state.edu/~gregoire/MIS/decide.htm (10 June 1999).

Guy, Mary E. 1990. "Decision Making." In *Ethical Decision Making in Everyday Work Situations*, 25–44. Westport: Greenwood Press.

———. 1990a. "Guidelines for Making Ethical Decisions." In *Ethical Decision Making in Everyday Work Situations*, 155–67. Westport: Greenwood Press.

Harmes, Harry H. 1994. "Genetic Decision Making Model." http://www.woodrow.org/teachers/bi/1994/genetic_decision_making.html (7 June 1999).

Harris, C. E., M. Pritchard, and M. Rabins. 1995. *Engineering Ethics: Concepts and Cases*. Belmont: Wadsworth.

Hiskes, Anne L., and Richard P. Hiskes. 1986. *Science, Technology, and Policy Decisions*. Boulder: Westview Press.

Holt Consulting Services. 1999. "Questions for Ethical Decision-Making." http://www.holtconsulting.com/decision.html (9 June 1999).

Hopper, Carolyn. 1999. "Top Ten Questions You Should Ask Yourself When Making an Ethical Decision." http://www.mtsu.edu/~u101irm/ethicques.html (10 June 1999).

Josephson Institute of Ethics. 1999. "Five Steps of Principled Reasoning." http://josephsoninstitute.org/MED/med5steps.htm (10 June 1999).

Kallman, Ernest A., and John P. Grillo. 1996. "Solving Ethical Dilemmas: A Sample Case Exercise." In *Ethical Decision Making and Information Technology*, 33–56. New York: McGraw-Hill.

Kidder, Rushworth M. 1995. *How Good People Make Tough Choices*. New York: William Morrow.

Kohlberg, Lawrence. 1981. *The Meaning and Measurement of Moral Development*. Cambridge: Oelgeschlager, Gunn and Hain.

Koop, Rebecca. 1998. "Ethical Analysis." http://www.coba.wright.edu/msis/mis323/slides/eth-plan/tsld005.htm (10 June 1999).

Kvanvig, Jonathan L. 1999. "The RESOLVEDD Method." http://kvanvig.tamu.edu/resolvedd.htm (10 June 1999).

Liffick, Blaise W. 1995. "Analyzing Ethical Scenarios." http://cs.millersv.edu/~liffick/scenario.html (6 Jan. 2002).

Long, Deborah H. 1998. "Models for Ethical Decision Making." In *Doing the Right Thing: A Real Estate Practitioner's Guide to Ethical Decision Making*, 39–54. 2d edition. Upper Saddle River: Prentice-Hall.

MacDonald, Chris. 1995. "A Guide to Moral Decision Making." http://www.ethics.ubc.ca/chrismac/publications/guide.html (7 Mar. 1998).

Maner, Walter. 1996. "Unique Ethical Problems in Information Technology." *Science and Engineering Ethics* 2: 137–54.

———. 1981. *Practicum Handbook*. New York: Helvetica Press.

Mathison, David L. 1987. "Teaching an Ethical Decision Model That Skips the Philosophers and Works for Real Business Students." In *Proceedings*, 1–9. New Orleans: National Academy of Management.

McDonald, Michael. 1998. "A Framework for Ethical Decision-Making, Version 4, Ethics Shareware." http://www.ethics.ubc.ca/mcdonald/decisions.html (7 June 1999).

McLaren, Ronald. 1989. *Solving Moral Problems: A Strategy for Practical Inquiry*. Mountain View: Mayfield.

Miller, Keith. 1989. "Paramedic Prose." Personal e-mail (19 June 1989).

Miller, Will. 1996. "Ethics and Decision-making: A Set of Questions." http://plsc.uark.edu/book/books/ethics/guide.html (14 June 1999).

Moulder, Linda. 1997. "Ethical Decisionmaking Process." http://www.biology.ewu.edu/students/Courses/bio340/decisionmaking.html (10 June 1999).

Nash, Laura L. 1981. "Ethics Without Sermons." In *Howard Business Review* 59: 79–90.

Newman, Dianna L., and Robert D. Brown. 1996. "A Framework for Making Ethical Decisions." In *Applied Ethics for Program Evaluation*, 91–120. Thousand Oaks: Sage.

Parker, Donn B. 1978. *Ethical Conflicts in Computer Science and Technology*. Menlo Park: SRI International.

Pollack, John L. 1998. "Procedural Epistemology." In *The Digital Phoenix: How Computers Are Changing Philosophy*, edited by Terrell Ward Bynum and James H. Moor, 17–36. Oxford: Blackwell.

Rananu, Harjinder, Jennifer Davies, and Simon Rogerson. "Ethical Analysis of Software Failure Cases." In *Proceedings of ETHICOMP 96*, edited by Porfirio Barroso, Terrell Ward Bynum, Simon Rogerson, and Luis Joyanes, 364–83. Madrid: Complutense University.

Rogers, William J. 1997. "Guidelines for Facilitating Solutions to Ethical Dilemmas." http://www.engr.washington.edu/~uw-epp/Pepl/Ethics/ethics4.html (9 June 1999).

———. 1997a. "Nine Basic Steps to Personal Ethical Decision Making." http://www.engr.washington.edu/~uw-epp/Pepl/Ethics/ethics4.html (9 June 1999).

Ruggiero, Vincent Ryan. 1992. "The Basic Criteria." In *Thinking Critically About Ethical Issues*, 53–67. 3d edition. Mountain View: Mayfield.

Schaefer, Arthur Gross. 1999. "A Suggested Strategy for Ethical Decision Making." In *The Jewish Ethics Challenge*. 1 Jan. 1999. http://www.syn2000.org/Articles/Schaefer1/node25.html (11 May 1999).

Schlossberger, Eugene. 1993. "Ethical Decision Making." In *The Ethical Engineer*, 23–38. Philadelphia: Temple University.

Searing, Donald R. 1998. *HARPS Ethical Analysis Methodology: Method Description*. Taknosys Software Corporation.

Sechowski, Melissa. 1998. "Introduction to Ethical Decision Making and Case Evaluations." http://oit.iusb.edu/~msechows/Ethics.html (10 June 1999).

Spiceland, J. David, and James F. Sepe. 1995. *Intermediate Accounting*. New York: McGraw-Hill.

Spinello, Richard A. 1997. "Frameworks for Ethical Analysis." In *Case Studies in Information and Computer Ethics*, 22–50. Upper Saddle River: Prentice-Hall.

Steinman, Sarah O., Nan Franks Richardson, and Tim McEnroe. 1998. "The Ethical Decision-Making Process." In *The Ethical Decision-Making Manual for Helping Professionals*, 17–22. New York: Brooks/Cole.

Taylor, Shaun N. 1999. "Bioethical Analysis Worksheet." http://www.accessexcellence.org/21st/TE/BE/worksheet.html (9 June 1999).

Thomasma, David C., and Patricia A. Marshall, eds. 1995. *Clinical Medical Ethics Cases and Readings*. Lantham: University Press of America.

United States Department of Defense. 1999. "Joint Ethics Regulation DoD 5500.7-R." http://www.defenselink.mil/dodgc/defense_ethics/ethics_regulation/jer1-4.doc (10 June 1999).

Velasquez, Manuel, Claire Andre, Thomas Shanks, and Michael J. Meyer. 1998. "A Framework for Ethical Decision Making." In *Practicing Ethics*. http:/www.scu.edu/SCU/Centers/Ethics/practicing/decision/thinkingshtml (6 Jan. 2002).

von Hooft, Stan, Lynn Gillam, and Margot Byrnes. 1995. "Ethical Decision Making." In *Facts and Values: An Introduction to Critical Thinking for Nurses*, 255–70. Philadelphia: MacLennan and Petty.

Waldfogel, Dean. 1996. "Ethical Decision-making Model." http://www.iusd.k12.ca.us/curriculum/ethics/tsld028.htm (9 June 1999).

Woodruff, Brian. 1992. "Presenting Ethical Dilemmas in the Classroom." http://www.accessexcellence.org/AE/AEPC/WWC/1992/dilemmas. html (9 June 1999).

16

LILLIPUTIAN COMPUTER ETHICS

JOHN WECKERT

Nanotechnology and quantum computing have the potential to radically change information technology. If research in these technologies is successful, and there are signs that it will be, computers will become very, very small and very, very fast, and will have an enormous amount of memory relative to computers of today. This possibility is creating excitement in some quarters, but anxiety in others. Speaking of nanotechnology, in a recent and much publicised article, Bill Joy wrote that

> ... it is most of all the power of destructive self-replication in genetics, nanotechnology, and robotics (GNR) that should give us pause. .... The only realistic alternative I see is relinquishment: to limit development of the technologies that are too dangerous by limiting our pursuit of certain kinds of knowledge. (Joy 2000)

Given the variety of benefits promised by nanotechnology in medicine, the environment, and information technology, to pick out just a few, Joy's claim seems a little strong. This essay will discuss a few potential developments to see what the appropriate reaction to this technology is. Are the worries enough to give his call for a limit on research any weight?

Before proceeding to examine the claim that some research in computing should not take place, we need to look at nanotechnology and quantum computing, to see what dangers they may pose.

Nanotechnology and Quantum Computing

Nanotechnology is relatively new, although it was first mooted by Richard Feynman in a lecture entitled "There's plenty of room at the bottom," in 1959. Nanotechnology involves the manipulation of individual atoms and molecules in order to build structures. One nanometre is one billionth of a metre (three to five atoms across), and "nanoscience and nanotechnology generally refer to the world as it works on the nanometer scale, say, from one nanometer to several hundred nanometers" (NSTC 1999). This tech-

This essay was originally presented as a paper at the ETHICOMP2001 conference in June 2001 in Gdansk, Poland.

nology is important because it enables very small things to be built and gives great control over the properties of the materials constructed. This technology is claimed to have enormous benefits in a variety of fields. Nanoparticles inserted into the body could diagnose and cure diseases, building materials could adapt to the weather conditions, cheap and clean energy could be produced, sensing devices could become much more sensitive, and computers could be made much faster and could have more memory. While these benefits have not yet accrued, they are not mere speculation. The theories underlying the proposed applications are, it is argued, scientifically based. The technology for manipulating atoms individually is available: In 1989 IBM physicists produced the IBM logo by manipulating atoms.

The main interest of this paper is the relationship of nanotechnology to computing. Quantum computers would be much more powerful than any available today. According to one researcher, it is expected that a quantum computer would be able to perform some computations that could not be performed by all the current computers on the planet linked together, before the end of the universe (Simmons 2000). Computers will also become very small. For example, at one end of the spectrum there would be devices that incorporated "nanoscale computers and several binding sites that are shaped to fit specific molecules, [that] would circulate freely throughout the body . . ." (Merkle 1997). At the other end "we should be able to build mass storage devices that can store more than 100 billion billion bytes in a volume the size of a sugar cube, and massively parallel computers of the same size that can deliver a billion billion instructions per second—a billion times more than today's desktop computers" (Merkle 1997). Smalley is a little more cautious, but still estimates that the efficiency of computers would be increased by a factor of millions (Smalley 1999). In addition, there could be "detecting devices so sensitive that they could pick up the equivalent of the drop of a pin on the other side of the world" (Davies 1996).

This technology, if it develops in the predicted manner, will facilitate some interesting developments. Monitoring and surveillance will become very easy, particularly when the new computer and communication technologies are linked. People with microscopic implants will be able to be tracked using Global Positioning Systems (GPS), just as cars can be now, only more efficiently. One need never be lost again! Other implants could increase memory, reasoning ability, sight, hearing, and so on. Some argue that the distinction between humans and machines might no longer be useful (Kurzweil 1999). And virtual reality could become indistinguishable from reality itself.

Before considering some of the ethical issues raised by these possible developments, it is worth asking just how credible are the claims. Apparently many are quite credible. Many are being made by researchers with impressive records in computer science and in other sciences, and research is being supported by reputable universities and governments. More important, some progress has been made. It is already possible to manipulate individual atoms, as stated earlier; nanotubes have been developed; and some progress has been made in building simple quantum computers. Theoretically, nanotechnology can work. There is enough evidence to suggest that nanotechnology and quantum computing will become realities, probably sometime around the middle of this century, and so it is worth looking at any ethical questions that they might raise.

Ethical Questions

While nanotechnology has potential benefits and dangers in a wide variety of areas—for example, in health and in the environment, as previously mentioned—we will consider here just some potential dangers in the field of computing. That there are also benefits in this area is not being questioned.

There are at least two sets of issues. One set concerns existing problems that will be exacerbated by the miniaturisation of computers. This miniaturisation will involve the development of smaller, much more powerful machines with much more sensitive input devices. The second set concerns potential problems, problems that as yet have not arisen, at least not in any significant way.

Exacerbated Problems

It is likely that most existing ethical problems in computing will be exacerbated. Easier and faster copying onto ever smaller devices will make protecting intellectual property more difficult. There will be more worries about Internet content as that content becomes more feasible and more difficult to control. It is likely, however, that one of the greatest impacts will be on privacy.

Privacy problems will be enormously increased. Vast databases that can be accessed at very high speeds will enable governments and businesses to collect, store, and access much more information about individuals than is possible today. In addition, the capacity for data mining, the exploration and analysis of very large amounts of data for the purpose of discovering meaningful and useful rules and patterns, will increase dramatically. And perhaps most important, the monitoring and surveillance of workers, prisoners, and even the general population will be greatly enhanced with the use of small, powerful computers and new sensing devices for input. GPS will be able to specify the location of individuals; cameras with artificial neural nets (or other learning technologies) will be able to pick out unusual behaviour in crowds or just on the streets and notify authorities.

While all of these possibilities have benefits (for example, for safety and efficiency), the possibility is also opened for large-scale control of individuals, either by governments, employers, or others with authority.

There will be a need for the reassessment of privacy legislation, the use of personal information by governments and corporations, and guidelines and legislation for the use of monitoring devices.

New Problems
Some problems are new in the sense that so far it has not been necessary to face them in any realistic way. Just three will be mentioned here.

Artificial intelligence. If machines are developed that behave in much the same way as humans do, in a wide variety of contexts, the issue of whether they are things with moral rights and responsibilities will arise. Consideration would need to be given to how they should be treated. If they behaved like us, would we be justified in treating them differently?

Bionic humans. Chip implants that enhance various senses, memory, and perhaps even other capacities such as reasoning ability and creativity may blur the distinction between human and machine. We already have eyeglasses, hearing aids, cochlear implants, hair implants, skin grafts, tooth implants, pacemakers, transplants, and so on, so why should more advanced implants matter? Perhaps there is a difference between helping people to be "normal"—that is, correcting a deficiency—and making a normal person a "super human." But just why this is the case would need to be spelt out.

Virtual reality. Virtual reality systems will improve to the point where it may become difficult to tell the difference between "real" and "virtual" reality. There may be no apparent difference between really hang gliding and doing so virtually.

Should the Research Be Controlled?

If Joy is right about the dangers of this new technology, then there is some research that computer scientists ought not do, or, if they do it, they ought to be held morally responsible for the consequences of that research. Or so it would seem. There are, however, a number of issues that need to be sorted out before we can be confident in affirming this. One concerns the differences between pure research and technological development; another issue is the use to which that development is put. The first question here is whether there is any pure research that should not be undertaken—that is, whether there is any "forbidden knowledge": Is there any knowledge that we should not attempt to discover? A related question is whether there is any technology that should not be developed, and there is the further question of limits to the uses of that technology.

We should bear in mind that there are distinctions between knowledge, the technology developed from that knowledge, and the uses to which the knowledge is put. Certain uses of knowledge (or the technology based on the knowledge) ought to be avoided if those uses cause harm. The emphasis for the moment is on the knowledge itself. The knowledge must also be distinguished from the method of gaining that knowledge. Clearly certain methods for gaining knowledge are wrong, for example, those that cause harm. (In particular situations some greater good may make some degree of harm, both in the gaining of knowledge and in its use, permissible.) The question of whether the knowledge itself ought to be forbidden is, or seems to be, quite a different matter. Knowledge is neither morally good nor morally bad in the way that methods or uses might be.

It is difficult to make sense of the claim that there is some knowledge that is morally bad in itself regardless of any consequences. It is certainly difficult to find any examples. It is easy to find examples of knowledge with harmful consequences. Joy himself seems to acknowledge this when he says that knowledge of nanotechnology should be limited in order to limit the technologies that would be developed from that knowledge. If knowledge has harmful consequences, should it be forbidden? Not always, because many types of knowledge can be put to both beneficial and harmful uses, and we do not automatically want to rule out the beneficial uses.

At this point it is worth looking briefly at Somerville's argument in *The Ethical Canary*, because perhaps the discussion should not be couched just in consequentialist terms. Her two basic principles are these:

> We have a profound respect for life, in particular human life . . . and we must act to protect the human spirit—the intangible, invisible, immeasurable reality that we need to find meaning in life and to make life worth living—that deeply intuitive sense of relatedness or connectedness to the world and the universe in which we live. (Somerville 2000, xi–xii)

She suggests that if scientific research violates either one of these principles, then it ought to be avoided even if it has some beneficial consequences. Her second principle has particular relevance for nanotechnology and quantum computing, although she does not discuss these fields. We shall touch on this principle later, and now consider some more consequentialist arguments.

To help to clarify matters, the consequences of knowledge can be divided into two groups: physical and mental. Physical consequences are always uses to which the knowledge is put. Of most concern is knowledge that will almost certainly be used in harmful ways. There may be no way to prevent the harm without preventing the knowledge in the first place. A case could be made that such knowledge is not the fit subject of research and ought to be forbidden.

Mental consequences do not necessarily involve uses. Some knowledge is such that simply knowing it has negative consequences. Nicholas Rescher puts it this way:

There are various things we simply ought not to know. If we did not have to live our lives amidst a fog of uncertainty about a whole range of matters that are actually of fundamental interest and importance to us, it would no longer be a human mode of existence that we would live. Instead, we would become a being of another sort, perhaps angelic, perhaps machine-like, but certainly not human. (Rescher 1987, 9)

Suppose that as a result of research in IT scientists learned how to build machines that in behaviour were indistinguishable from humans, and moreover, that it was obvious that these machines were purely deterministic and without free will. If we knew this, we would obviously have to see ourselves in a new light. Would we, in our present stage of evolution, be able to cope? If the GNR technology discussed by Joy developed in the manner that he fears, would we be able to continue to live happy and satisfying lives? If some knowledge has profound effects on the way we see ourselves, should it be forbidden? It seems that here, just as in the case of knowledge that almost inevitably leads to harm, a plausible argument can be made for forbidding it.

It has just been suggested that a case can be made in certain circumstances to restrict or prohibit research, but who should do this restricting or prohibiting? Does the state have a role? A strong argument for the freedom of science from political control is supplied by David Baltimore:

First, the criteria determining what areas to restrain inevitably express certain sociopolitical attitudes that reflect a dominant ideology. . . . Second, attempts to restrain directions of scientific inquiry are more likely to be generally disruptive of science than to provide the desired specific restraints. (Baltimore 1979, 41)

A number of arguments are offered to support these claims. First is what Baltimore calls the "Error of Futurism," that is, the supposition that we can predict the consequences of any research accurately enough to make any sensible decisions. The second argument is a version of one of John Stuart Mill's. Freedom of speech and expression allows the development of new ideas, increases the choices in life, and generally renews and vitalises life and makes it richer. A third argument is that repression in scientific research is likely to lead to repression in other areas and so will increase fear rather than strengthen society. A fourth argument is based on the unpredictability of science. Even if some research is not allowed, the knowledge to which it may have led might emerge from other research quite unexpectedly.

These arguments are aimed at pure or basic research, and Baltimore admits that the further one moves toward applications of research, the weaker these arguments become. If these arguments hold for pure research, then all of the responsibility for undertaking worthwhile research rests on the scientists themselves, which is where, according to Baltimore and others, it ought to rest. It must be noted here that it does not necessarily follow that the scientists' responsibilities extend to the *uses* to which the knowledge is put. They create or generate the knowledge from their research, but others decide how the knowledge is to be used. Whether this separation of responsibilities is ultimately sustainable is another matter, but for the moment we shall accept it.

We now return to Baltimore's arguments that research should not be externally controlled. His first argument, the "Error of Futurism," is that prediction is too unreliable to provide the basis for any restrictions on research. Consider, for example, Weizenbaum's prediction that research into speech recognition could have no useful consequences (Weizenbaum 1984, 271). It now appears to be an important tool in human-computer interface design for users with certain disabilities. Prediction is certainly fraught with danger; however, we often must base our actions on predictions, on what we believe may happen, and it is not clear why the case of research should be any different.

Baltimore's second argument is that freedom of speech and expression allows the development of new ideas, thereby increasing life's choices and generally making it richer. This is true, and this form of freedom is undoubtedly an important good, but it is not the only one, and it can be in conflict with others. In general we are restricted in the performance of actions that will, or are likely to, harm others. Again, it is unclear why research should be treated differently.

The third argument is that repression in scientific research is likely to lead to repression in other areas and so will increase fear rather than strengthen society. While we can agree that repression is not good, many things are restricted or repressed in a civilised society—for example, certain research using human subjects and driving under the influence of alcohol—but these restrictions surely reduce rather than increase fear. If probable harm is as closely associated with knowledge as suggested earlier, there is no reason why pure research should be treated differently from other aspects of life. Baltimore's final argument is that, because of the unpredictability of science, the undesired knowledge might emerge unexpectedly from research that was not disallowed. This is true but not to the point. While it may not be possible to ensure that some undesirable knowledge will not be discovered, it is almost certainly possible to reduce the probability that it will be.

It is, then, permissible or even obligatory on occasions to restrict or forbid research on the ground that mental or physical harm is likely to result from it. However, this should not be done lightly, because freedom in this area is important, not only for the good of science but also for the good of society. There should be a presumption in favour of freedom of research. If research is to be restricted, the burden of proof should be on those who want to restrict it. However, there seems to be a conflicting intu-

ition here. If a *prima facie* case can be made that some particular research
will most likely cause harm, either mentally or physically, then the burden
of proof should be on those who want the research carried out to demonstrate that it is safe. The situation then appears to be this. There is a
presumption in favour of freedom until such time as a *prima facie* case is
made that the research is dangerous. The burden of proof then shifts from
those opposing the research to those supporting it. At that stage the
research should not begin or be continued until a good case has been made
that it is safe.

The conclusion to this point, then, is that the case against the state's
having a role in the control of scientific research has not been made, but
that such control has dangers and it should not be embraced lightly. The
argument so far has focussed primarily on pure research, because that is
where it is most difficult to make a case for control. However, the case of
technological development, one of the fruits of pure research, is not obviously much different. Just as a scientist can say that he or she is just adding
to knowledge and therefore has no responsibility for the use to which that
knowledge is put, so technologists can say that they are just developing
tools, and it is not up to them how those tools are used.

Nanotechnology and Quantum Computing Research

It has been argued in this essay that nanotechnology and quantum computing do raise some worrying ethical questions. While these technologies
offer the potential for great benefit, it is not benefit unalloyed, and some of
the potential problems were outlined. It has also been argued that there are
cases in which halting certain types of research could be justified. The
question here is whether research into nanotechnology and quantum
computing is in this category.

Given the quite fundamental changes that these technologies could
facilitate, it is not enough merely to consider the potential benefits and
harms as they might apply to life as we know it now. The issue is more one
of the kind of life that we want. Can we, and do we want to, live with artificial *intelligences*? We can happily live with fish that swim better than we
do, with dogs that hear better, hawks that see and fly better, and so on, but
things that can reason better seem to be in a different and altogether more
worrying category. Do we want to be "super human" relative to our current
abilities, with implants that enhance our senses, our memories, and our
reasoning ability? What would such implants do to our view of what it is
to be human? Does it matter if our experiences are "real" or not, that is, if
we have them in a virtual world or in the real one? Would there be any
sense in that distinction? These are all big questions that cannot be
answered here, but the suggestion is that they are the important ones to
address when considering the future of research into nanotechnology and
quantum computing. These questions seem related to Somerville's second

principle, that of protecting the human spirit. Perhaps research into nanotechnology and quantum computing does not violate that principle, but much more examination of the issues is warranted.

References

Baltimore, David. 1979. "Limiting Science: A Biologist's Perspective." In *Limits of Scientific Inquiry*, edited by G. Holton and R. S. Morrison, 37–45. New York: W. W. Norton.

Davies, Paul. 1996. "Foreword." In *Quantum Technology*, edited by Gerard Milburn. London: Allen and Unwin.

Feynman, Richard. 1959. "There's Plenty of Room at the Bottom: An Invitation to Enter a New Field of Physics." A talk given on December 29 at the annual meeting of the American Physical Society at the California Institute of Technology. Available online at http://www.zyvex.com/nanotech/feynman.html. Accessed March 27, 2001.

Joy, Bill. 2000. "Why the Future Doesn't Need Us." *Wired*, 8.04.

Kurzweil, Ray. 1999. *The Age of Spiritual Machines: When Computers Exceed Human Intelligence*. London: Allen and Unwin.

Merkle, Ralph. 1997. "It's a Small, Small, Small World." *MIT Technology Review* (Feb./Mar.): 25. Available online at http://www.techreview.com/articles/fm97/merkle.html. Accessed November 2, 2000.

NSTC. 1999. "Nanotechnology: Shaping the World Atom by Atom." A report of the National Science and Technology Council (NSTC), Committee on Technology, The Interagency Working Group on Nanoscience, Engineering and Technology. Available online at http://itri.loyola.edu/nano/ IWGN.Public.Brochure/. Accessed March 27, 2001.

Rescher, Nicholas. 1987. *Forbidden Knowledge and Other Essays on the Philosophy of Cognition*. Dordrecht: D. Reidel.

Simmons, Michelle. 2000. "Faster, Smaller, Smarter." A talk given at the forum Small Things, Big Science: Nanotechnology, Horizons of Science. University of Technology, Sydney, November 23, 2000.

Smalley, R. E. 1999. "Nanotechnology." Prepared written statement and supplemental material of R. E. Smalley, Rice University, June 22. Available online at http://www.house.gov/science smalley_062299.htm. Accessed March 27, 2001.

Somerville, Margaret. 2000. *The Ethical Canary: Science, Society and the Human Spirit*. New York: Viking.

Weizenbaum, Joseph. 1984. *Computer Power and Human Reason: From Judgment to Calculation*. New York: Penguin Books. Originally published in San Francisco by W. H. Freeman, 1976.

17

DEONTIC LOGIC AND COMPUTER-SUPPORTED COMPUTER ETHICS

JEROEN VAN DEN HOVEN and GERT-JAN LOKHORST

In the early 1980s philosophers, computer scientists, and legal scholars began to think systematically about the ethical issues in computing (Johnson 1985; Moor 1985; Bynum 1985). Most of the issues that were discussed at that time are still on today's research agenda: privacy and data protection, intellectual property in information, responsibility for design and use of information systems, equal access to information.

In addition to the more traditional methods of moral inquiry into these issues, several attempts have been made to utilize computer programs and information systems to support moral reasoning and help us understand moral behavior. Should these attempts be successful, we would be presented with an extraordinary full circle: computer technology would come to the aid of those grappling with the moral problem to which computer technology itself has given rise – computer-supported computer ethics. Different research projects along these lines can be distinguished. First, computer-assisted game theory (Danielson 1992), applied cognitive science (Goldman 1993), and AI research (see, for example, Castelfranchi 2000) help us to get a better understanding of moral behavior, its origin, dynamics, and rationality. Second, computer-supported checklists and decision support systems assist moral decision makers in difficult cases in and outside the field of IT (Gotterbarn and Rogerson 1999). Third, multi-media tools (Cavalier) help us to study and teach real-life cases. Finally, computer systems have been used to implement and execute deontic reasoning (Lee 1992) both in law and in commerce. Deontic models and computer programs may help us to handle the many deontic constraints associated with electronic contracting (buying and selling, promising and authenticating documents) in eCommerce and eBusiness (Tan 2000).

The ethical issues in computing and information and communication technology seem to have little in common. They range from the desirability of software patents to the acceptability of the Communications Decency Act, from the need for genetic privacy to the prospects of cyber-democracy, and from identity theft to the selection of new top-level domain names. There are some commonalities, however, in the moral language used to articulate the moral problems of an information society and to talk *about* them. In this essay we, first, provide a description and informal analysis of the commonalities in moral discourse concerning issues in the field of information and communications technology. Second, we present a logic model (DEAL) of this type of moral discourse that makes use of recent research in deontic, epistemic, and action logic. Third, we indicate – drawing upon recent research in computer implementations of modal logic – how information systems may be developed that implement the proposed formalization.

1. Ethical Issues and Information Technology

1.1. Intellectual Property

The central question in the field of intellectual property (IP) is concerned with the *justification* of IP rights – in particular, "Why should there be IP rights at all, how can we establish their scope and argue for their application in particular cases? How does ownership of information (and software) justify the rights holder to limit the freedom of others to use the information (software) concerned?" Another part of the IP debate is concerned with more practical questions, such as "How are IP rights adequately expressed in laws, regulations, and social institutions?"

Property rights in information constitute moral constraints on the actions of others vis-à-vis the protected information.

(A) If John has an IP right in a particular piece of information X, then Peter ought to have permission from John to acquire, process, or disseminate X.

1.2. Privacy and Data Protection

The privacy and data-protection debate is concerned with the justification of claims to limit access to personal information. There are different moral grounds on which one can argue that constraints should be placed on the dissemination, processing, and acquisition of personal information. One way to characterize data-protection rights is to say that they are moral constraints on what persons may do with one's personal information.

(B) If information X is about John and if Peter does not have X, then Peter is not permitted to acquire X without John's consent. If he does have X, then he is not permitted to process or disseminate it without John's consent.

Peter ought to have John's permission if he wants to acquire, process, or disseminate X. He is not free to provide X to others. So he is not permitted to inform himself or others about X, where X is information about a person P, without P's consent.

1.3. Equal Access

The central question in the debates usually lumped together under the heading of 'the divide between the information haves and the information have-nots', or 'the digital divide', is that some (types of) information X are so important for individuals that some persons or agencies have an obligation to see to it that individuals are treated equally so far as the availability of and access to X is concerned. Access to X ought to be distributed fairly. This would imply obligations on the part of government, for example, to supply X to all citizens and to remove impediments that may keep individual citizens from getting X, or it may be the case that if someone has a right to know something, then all have a right to know it – equal access.

(C) If A is informed about X, then all ought to be informed about X.

1.4. Responsibility and Information
Information technology provides us with tools to process information, to acquire knowledge, and to make data available. As with other technologies, the fact that it broadens the range of our actions gives rise to moral questions. Debates about reproductive and nuclear technology revolve around the question of whether we should do what we are technically able to do in these fields. Information technology draws our attention to moral questions at the intersection of agency, morality, and epistemology. Are we responsible for what we and others (do not) know or believe, and are we also responsible for the design of our electronic epistemic artifacts and the software that functions effectively as a doxastic policy? Do we have responsibilities to make others believe certain things in certain circumstances? Is it permissible for our actions to affect our knowledge base and that of others in such a way that we can no longer be held accountable for what we do or think?

(D) If John has an information responsibility regarding X, then John has an obligation to see to it that specific others have access to information X.

1.5. General Form of Ethical Statements Concerning Information Acts
Consider the following sentences:

1. A informs B
2. A tells B that p
3. A lets B know that p
4. A shares his knowledge that p with B
5. A informs B about p
6. A sends a message to the effect that p to B
7. A's communications to B indicate that p

The general form of (1) to (7) can be rendered as:

- Agent A in informational context C sees to it that Agent B believes that p, or
- A informs B that X

Moral or legal constraints in information contexts may be expressed in general terms as follows:

- It is (not) obligatory or (not) permitted for A to see to it that B knows that p

There are three conceptual ingredients in this type of statement:

Deontic	Action	Epistemic/Doxastic
The right	to get	information
The obligation	to see to it that	others know
The permission	to let someone	know
Duty	to prevent people from	believing falsehoods
The right	to remain	ignorant

The vocabulary used or needed to capture moral talk about actions with respect to information thus comprises:

1. Agents
2. Information contexts
3. Information acts (tokens)
4. Information actions (types: acquisition, processing, dissemination)
5. Informational content (propositions)
6. Information relations between agents
7. Deontic constraints on (information actions of agents standing in) information relations
8. Revealed or tacit moral justifications for deontic constraints
9. Deontic operators (obligation and permission) capturing deontic constraints
10. Epistemic and doxastic operators (knowledge and belief) capturing cognitive states of agents
11. Action operators (sees to it that) capturing the actions of agents

In the next section, we sketch to what extent these notions have been studied in logical terms.

2. DEAL (Deontic/Epistemic/Action Logic)

There are three classes of notion that play an essential role in the type of discourse in which we are interested here: deontic notions, epistemic notions, and notions having to do with action. All three classes of notion have been intensively studied in philosophical logic.

2.1. Deontic Logic

Deontic logic studies the logic of obligation, permission, and prohibition. It is seventy-five years old and has been extensively applied in computer science (Meyer and Wieringa 1993). Deontic logic has one basic operator: O ("it is obligatory that"). O transforms a well-formed formula A into another well-formed formula, OA. For example: if A stands for "John stops for the red traffic light," then OA stands for "it is obligatory that John stops for the red traffic light."

Several other deontic notions can easily be defined in terms of O: PA ("it is permitted that A") $= \neg O \neg A$, FA ("it is forbidden that A") $= O \neg A$.

Standard deontic logic has the following axioms and rules of inference:

1. All classical tautologies
2. $O(A \rightarrow B) \rightarrow (OA \rightarrow OB)$
3. $OA \rightarrow PA$ (obligation implies permission)
4. If A and $A \rightarrow B$ are theorems, then so is B (modus ponens)
5. If A is a theorem, OA is a theorem

Standard deontic logic is a branch of modal logic and has the same kind of semantics (that is, Kripke-style semantics, characterized by accessibility relations between possible worlds).

2.2. Epistemic Logic

Epistemic logic is the logic of statements about knowledge and belief. It is about forty years old and has been extensively applied in computer science (Meyer and Van den Hoek 1995; Fagin, Halpern, Moses, and Vardi 1995).

Epistemic logic has two basic operators, which cannot be defined in terms of one another, namely,

- K_a ("agent a knows that")
- B_a ("agent a believes that")

Like O, K_a and B_a transform well-formed formulas into well-formed formulas. For example: suppose again that A stands for "John stops for the red traffic light" – then $K_a A$ stands for "agent a knows that John stops for the red traffic light," whereas $B_a A$ stands for "agent a believes that John stops for the red traffic light."

Standard epistemic logic has the following axioms and rules of inference:

1. All classical tautologies
2. $K_a (A \rightarrow B) \rightarrow (K_a A \rightarrow K_a B)$
3. $B_a (A \rightarrow B) \rightarrow (B_a A \rightarrow B_a B)$
4. $K_a A \rightarrow B_a A$ (knowledge implies belief)
5. $K_a A \rightarrow A$ (knowledge presupposes truth)
6. $K_a A \rightarrow K_a K_a A$, $K_a A \rightarrow K_a \neg K_a \neg A$ (optional)
7. $B_a A \rightarrow B_a B_a A$, $B_a A \rightarrow B_a \neg B_a \neg A$ (optional)
8. Modus ponens (as above)
9. If A is a theorem, $K_a A$ is a theorem

Standard epistemic logic is a branch of modal logic and has the same kind of semantics (that is, Kripke-style semantics, characterized by accessibility relations between possible worlds).

2.3. Logic of Action

The logic of action is concerned with the logical properties of statements about action. This branch of logic is at least fifty years old, but its most interesting developments have occurred comparatively recently (Belnap, Perloff, and Xu 2001). The logic of action has been used in computer science (for example, in dynamic logic), but the most recent developments in philosophical logic have not yet been applied in this field.

The basic operator of the logic of action as studied by Belnap, Perloff, and Xu (2001) is STIT ("sees to it that"). STIT is an operator that transforms a term a and a well-formed formula A into a well-formed formula [a STIT: A]. An example: if a is a term (denoting an agent a) and A stands for "the light is on," then [a STIT: A] stands for "agent a sees to it that the light is on" ("a switches the light on").

The logic of STIT may be axiomatized as follows. We only consider the single-agent case without so-called busy beavers. Definitions: $A^a = A \wedge \neg[a$ STIT: $A]$, $T = A \vee \neg A$.

1. All classical tautologies
2. $\neg[a$ STIT: $T]$
3. [a STIT: A] $\rightarrow A$
4. [a STIT: A] $\rightarrow$ [a STIT: [a STIT: A]]
5. [a STIT: A] $\wedge$ [a STIT: B] $\rightarrow$ [a STIT: $A \wedge B$]
6. [a STIT: [a STIT: A] $\wedge B$] $\rightarrow$ [a STIT: $A \wedge B$]
7. [a STIT: $A \wedge B$] $\wedge \neg[a$ STIT: B] $\rightarrow$ [a STIT: $A \wedge B^a$]
8. [a STIT: $\neg[a$ STIT: $A \wedge B] \wedge B^a$] $\rightarrow$ [a STIT: $\neg[a$ STIT: A] $\wedge B^a$]
9. [a STIT: A] $\leftrightarrow$ [a STIT: $A \wedge B^a$] $\vee$ [a STIT: $A \wedge \neg[a$ STIT: $A \wedge B^a$]]
10. [a STIT: $\neg[a$ STIT: $A \wedge [a$ STIT: $B \wedge \neg[a$ STIT: $B \wedge C^a$]]] $\wedge C^a$] $\rightarrow$ [a STIT: B]
11. [a STIT: A] $\leftrightarrow$ [a STIT: $\neg[a$ STIT: $\neg[a$ STIT: A]]]

12. Modus ponens (as above)
13. If $A \leftrightarrow B$ is a theorem, then [a STIT: A] $\leftrightarrow$ [a STIT: B] is a theorem

Extensions to multiple agents (joint agency) have also been studied. Postulate $a \neq b \rightarrow \neg$[a STIT: [b STIT: A]] is especially interesting in this context: an agent a cannot see to it that some different agent b sees to it that A (if a wants A to be the case, he must take care of this himself).

The logic of STIT is surprisingly rich, especially when combined with temporal notions. Some examples (Belnap, Perloff, and Xu 2001, ch. 9):

1. Could-have [a STIT: Q] is not equivalent to Might-have-been: [a STIT: Q]
2. If yon fellow sees to some state of affairs, then it might have been that the state of affairs did not obtain – at that very instant.
3. If a does something, then it might have been otherwise; that is, a might not have done it.
4. There is no reading of "The fact that a person could not have avoided doing something is a sufficient condition of his having done it" on which this claim is both interesting and true.
5. Invalid: "That we are responsible for some state of affairs implies that it must have been possible for us to have been responsible for its absence."
6. Invalid: "If a saw to something, then a could have refrained from seeing to it."
7. Suppose that a sees to it that Q; does it follow that a might have refrained from seeing to it that Q in the sense that there is a co-instantial alternative at which a refrains from seeing to it that Q? (STIT version: does [a STIT: Q] imply Might-have-been: [a STIT: $\neg$[a STIT: Q]]?) Answer: This implication is valid if and only if there are no "busy choosers."

The logic of STIT is a branch of modal logic (although STIT is a peculiar 'antinormal' operator). The semantics are similar to those of standard modal and temporal logic (possible worlds with certain relations between these worlds).

2.4. Hybrid Systems
When one wants to formalize the types of expressions mentioned in the previous section in terms of the operators we have mentioned, one quickly runs into "mixed" expressions, containing operators from more than one domain. Some examples:

1. $O (K_a A \rightarrow \forall x K_x A)$
   "it ought to be the case that everybody knows what a knows" – see (C) in sec. 1.3 above.

2. $O [a$ STIT: $\forall x K_x A]$
"a ought to see to it that everybody knows that A," that is, "a has an information responsibility regarding A" – see (D) in sec. 1.4 above.
3. $[a$ STIT: $\forall x (Fx \rightarrow B_x O A)]$
"a sees to it that everybody who is F believes that A is obligatory."
4. $\neg(P [a$ STIT: $A] \rightarrow \forall x P [x$ STIT: $A])$
Quod licet Jovi non licet bovi.

Such "mixed" expressions have not yet been very well studied. An exception is the joint logic of STIT and O (Belnap, Perloff, and Xu 2001, part 4). The STIT theorists view the STIT operator as particularly important in deontic contexts because they claim that deontic statements are usually of the form $O [a$ STIT $A]$. In other words, they maintain that such statements usually involve the deontic status of *actions* rather than *states of affairs*. (This is the "*Tunsollen* rather than *Seinsollen*" thesis from classical ethical theory.)

Combinations of STIT and epistemic operators are only briefly hinted at by Belnap, Perloff, and Xu (2001). Combinations of all three operators are not considered at all. Yet it will be clear that when one wants to express the views of computer ethicists in formally explicable terms, all these three types of operator are needed – and possibly even more. Interesting interactions between the operators might turn out to arise in the full-fledged system, and it hardly needs emphasizing that more work in this intriguing area is desirable.

3. Implementability

Trying to express one's views in logical terms is worthwhile in any case because it inevitably leads to more clarity than can otherwise be obtained. But trying to express one's views about computer ethics in terms of deontic, epistemic, and action logic is particularly attractive because the resulting theories are in principle implementable in computer software. As a result, one can partially relegate one's reasoning to the very machine about which one happens to be theorizing – the computer.

This is especially important when one is reasoning about the deontic and epistemic kinematics of large organizations employing hundreds of employees and serving thousands of clients or customers, each having privileges, responsibilities, duties, obligations, permissions, sources of information and misinformation, abilities and inabilities, and so on. In such circumstances, 'manual' reasoning quickly gets overwhelmed, and computer assistance becomes desirable. How should one delegate responsibilities, safeguard the flow of sensitive information, protect privacy, and so on, in today's complex organizational environments? Reasoning about such issues may be trivial so long as one is looking only at the level of individual agents, but the totality may be of mind-boggling complexity. It is precisely in problems of this type that the computer has traditionally come to the rescue.

What are the concrete prospects in this area? As we have said, all theories we have considered belong to the field of modal logic. So the question boils down to what extent modal logic is implementable. The answer is that the situation is similar to the situation in first-order predicate logic. This calculus is not fully implementable, but thanks to approaches like those embodied in PROLOG one can go surprisingly far – as far, in general, as needed for practical purposes. The situation in modal logic is beginning to look similar. Much work on the implementability of modal logic has been carried out during the past few years, and more progress has been achieved than one would have thought possible a decade ago. A detailed description of the results would go far beyond the scope of this essay; we refer interested readers to a site on the World Wide Web (Schmidt 2001) for a survey of recent achievements, especially with respect to modal theorem provers.

There is no denying that the software emerging from this field is still in its infancy, no matter how impressive the theoretical background may be. The existing programs are extremely unfriendly to the user, and they compile and run only under one or two variants of Unix. So far as the general user is concerned, the field is about as appealing as the Internet was in, say, 1985. But the potential benefits are great. More work in this area seems worthwhile.

4. Conclusion

A recent report from the Dutch Data Protection Authority ends on the following note:

> We conclude with an important piece of wisdom from the cypherpunks. The cypherpunks' credo can be roughly paraphrased as "privacy through technology, not through legislation." If we can guarantee privacy protection through the laws of mathematics rather than the laws of men and whims of bureaucrats, then we will have made an important contribution to society. It is this vision which guides and motivates our approach. (Hes and Borking 1998)

We are motivated by the same vision. But there is an important difference between our approach and the cypherpunks' position. They did not indicate at all how their goal might be achieved, whereas we have a clear view as to how one should proceed. By combining the most sophisticated computer ethics with the most advanced computer programs, based on the most solid results from philosophical logic, the cypherpunks' vision may well come within reach. We still have a long way to go, but there are no insurmountable obstacles on the horizon.

References

Belnap, Nuel, Michael Perloff, and Ming Xu. 2001. *Facing the Future: Agents and Choices in Our Indeterminist World.* New York: Oxford University Press.

Bynum, Terrell Ward, editor. 1985. *Computers and Ethics.* Special issue of *Metaphilosophy* 16, no. 4 (October).

Castelfranchi, Cristiano. 2000. "Artificial Liars: Why Computers Will (Necessarily) Deceive Us and Each Other." *Ethics and Information Technology* 2, no. 2: 113–19.

Cavalier, Robert. *Multimedia and Teaching Ethics.* Available online at http://www.andrew.cmu.edu/user/rc2z/profile.html. No date given.

Danielson, Peter. 1992. *Artificial Morality: Virtuous Robots for Virtual Games.* London: Routledge.

Fagin, R., J. Y. Halpern, Y. Moses, and M. Y. Vardi. 1995. *Reasoning about Knowledge.* Cambridge, Mass.: MIT Press.

Goldman, Alvin. 1993. "Ethics and Cognitive Science." *Ethics* 103: 337–60.

Gotterbarn, Don, and Simon Rogerson. 1999. "An Ethical Decision Support Tool: Improving the Identification and Response to the Ethical Dimensions of Software Projects." *Proceedings of ETHICOMP99 (Rome).* CD-ROM. Leicester: DeMontfort University.

Hes, Ronald, and John Borking, editors. 1998. *Privacy Enhancing Technologies: The Path to Anonimity.* The Hague: Registratiekamer. Achtergrondstudies en Verkenningen series, vol. 11.

Johnson, Deborah. 1985. *Computer Ethics.* New York: Prentice-Hall.

Lee, Ron. 1992. *DX: A Deontic Expert System Shell.* EURIDIS internal report 92.10.01b. Rotterdam: Erasmus University.

Meyer, John-Jules Ch., and Wiebe van der Hoek. 1995. *Epistemic Logic for AI and Computer Science.* Cambridge: Cambridge University Press.

Meyer, John-Jules Ch., and Roel J. Wieringa, editors. 1993. *Deontic Logic in Computer Science.* Chichester: John Wiley.

Moor, Jim. 1985. "What Is Computer Ethics?" In Bynum 1985, 266–76.

Schmidt, Renate. 2001. *Advances in Modal Logic.* Available online at http://www.cs.man.ac.uk/~schmidt/tools/.

Tan, Yao-Hua. 2000. "A Logical Model of Trust in Electronic Commerce." *Electronic Markets* 10: 258–63.

NOTES ON CONTRIBUTORS

Colin Allen is professor of philosophy at Texas A & M University, specializing in animal cognition, philosophy of mind, and philosophy of biology. His first piece of Internet software, *The Logic Daemon*, came online in 1992. Because it predated the World Wide Web, it initially communicated only via electronic mail but since 1995 has been resident at <http://logic.tamu.edu/>. He programs in Lisp, Perl, and C and started programming for the *Stanford Encyclopedia of Philosophy* in 1998, becoming associate editor in 1999. He is also the author of two articles for the *Encyclopedia*, on animal consciousness and on teleological notions in biology.

John A. Barker is professor emeritus of philosophy at Southern Illinois University, Edwardsville. He has written many articles on epistemology, logic, the philosophy of language, and the philosophy of mind. He is the author of *A Formal Analysis of Conditionals* and *ProtoThinker: A Model of the Mind*.

Anthony F. Beavers is associate professor of philosophy at the University of Evansville, where he directs the Internet Applications Laboratory. His research employs the study of metaphysics to develop search and index algorithms for organizing information. His computer projects include *Hippias: Limited Area Search of Philosophy on the Internet* (1997) and *Noesis: Philosophical Research Online* (1998).

Terrell Ward Bynum is professor of philosophy at Southern Connecticut State University and director of the Research Center on Computing and Society there. He is a lifetime member of Computer Professionals for Social Responsibility, past chair of the Committee on Professional Ethics of the Association for Computing Machinery, and past chair of the Committee on Philosophy and Computers of the American Philosophical Association. For twenty-five years he was editor in chief of *Metaphilosophy*. He is author and editor of numerous books, articles, and video programs on computer ethics, logic, psychology, education, and history of philosophy. He is coeditor, with James H. Moor, of *The Digital Phoenix: How Computers Are Changing Philosophy* (1998, revised 2000) and coeditor, with Simon Rogerson, of *Computer Ethics and Professional Responsibility* (2002).

Marvin Croy is associate professor and information technology coordinator in the Department of Philosophy at the University of North Carolina at Charlotte. His areas of research encompass the development of instructional technology, ethical questions concerning the appropriate use of such technology, issues in logic and problem solving, and theories of technology development. His work in these areas has been published in numerous journals and has been supported by both the National Science Foundation and the National Endowment for the Humanities.

Randall R. Dipert is C. S. Peirce Professor of American Philosophy at the SUNY University Center at Buffalo. He is the author of *Artifacts, Art Works, and Agency* (1993) as well as more than forty articles and reviews on logic, the history of logic, American philosophy, aesthetics, metaphysics, political philosophy, philosophy of science, action theory, and music theory.

Charles Ess is professor of philosophy and religion and director of the Center for Interdisciplinary Studies at Drury University. He has received awards for teaching excellence, scholarship, and his work in hypermedia. He has published in interdisciplinary and comparative ethics, computers and democratization, history of philosophy, Continental philosophy, and biblical studies. He has lectured nationally and abroad, and edited *Philosophical Perspectives on Computer-Mediated Communication* (1996) and *Culture, Technology, Communication: Towards an Intercultural Global Village* (2001).

Luciano Floridi (Faculty of Philosophy, Sub-faculty of Computation, Programme in Comparative Media Law and Policy, University of Oxford) has held fellowships and visiting professorships in Germany, the United Kingdom, and Italy. He is the author of more than thirty articles on the philosophy of computing and information and of *Philosophy and Computing* (1999), *Internet* (1997), and *Scepticism and the Foundation of Epistemology* (1996). He is currently editing the *Blackwell Guide to the Philosophy of Computing and Information* and was the consultant editor for the *Iter Italicum* on CD-ROM (1995) and the *Routledge Encyclopedia of Philosophy* on CD-ROM (1998). He is executive director of the Italian Web Site for Philosophy (www.swif.uniba.it).

Patrick Grim is SUNY Distinguished Teaching Professor at the State University of New York at Stony Brook, working with the informal Group for Logic and Formal Semantics in the Department of Philosophy. His work in logic and computational modeling has appeared in major journals of theoretical biology, evolutionary linguistics, computer graphics, and decision theory as well as journals in philosophy and philosophical logic. His books include *The Incomplete Universe: Totality, Knowledge, and Truth* (1991) and *The Philosophical Computer: Exploratory Essays in Philosophical Computer Modeling* (with Gary Mar and Paul St. Denis, 1998). He is also founding editor of the twenty-three volumes of *The Philosopher's Annual*.

Gene Korienek has considerable research and development experience with control systems and artificial intelligence projects in academia and industry, and as a consulting scientist. He was a senior research scientist at Johnson Controls, Inc. from 1983 to 1989, where he investigated self-organizing control strategies and adaptive control heuristics in the Artificial Intelligence Group of the Corporate Research Department. From 1989 to 1995 he owned and operated the artificial intelligence consulting company ARTIFACT, Inc., where he consulted extensively on object-oriented software design and development with Fortune 100 companies. He received his Ph.D. from Florida State University in 1992 and has since directed the Biological Robotics Project at Simon Fraser University and the Biological Control Lab at Oregon State University. He is also a NASA Ames Research Associate involved with the Mission to Mars project.

Gert-Jan C. Lokhorst has an M.Sc. degree in medicine (1980) and a Ph.D. degree in philosophy (1992). He is currently a lecturer in (a) logic and (b) the philosophy of artificial intelligence, in the Department of Philosophy at Erasmus University, Rotterdam. He has published many papers, and a book on consciousness and the brain (1986, in Dutch). More than a hundred of his publications are available at http://www.eur.nl/fw/staff/lokhorst/.

Pete Mandik is assistant professor of philosophy at William Paterson University and a member of the McDonnell Project in Philosophy and the Neurosciences at Simon Fraser University. He works in the areas of philosophy of mind, philosophy of cognitive science, and neurophilosophy. He is a coeditor of *Philosophy and the Neurosciences: A Reader* (2001) and a coauthor of *Principles of Cognitive Science* (forthcoming).

Walter Maner is professor of computer science at Bowling Green State University, Ohio. He has been a Woodrow Wilson Fellow, an NDEA Fellow, a Fulbright Scholar, and a Mellon Scholar-in-Residence. In 1976, he developed and taught the first university course entirely devoted to issues in computer ethics. In 1978, he delivered the first taxonomy for ethical issues in computing, at the Second National Workshop Conference on the Teaching of Philosophy, and the first model curriculum for courses in computer ethics, at the Computers and Society/Computer Impact Workshop in Williamsburg, Virginia. Between 1978 and 1985, the curriculum design was distributed as a monograph by the National Resource Center for the Teaching of Philosophy after being published by Helvetia Press and has since been republished in various journals. In 1979, he and Frank Marsh organized the Conference on the Computer as Physician, the first conference entirely devoted to ethical issues in computing. In 1991, he and Terrell Ward Bynum organized the Conference on Computing and

Values, the first international conference entirely focused on computer ethics and the first in the ongoing ETHICOMP series.

James H. Moor is professor of philosophy at Dartmouth College. He has written numerous research articles in computer ethics, philosophy of artificial intelligence, philosophy of mind, philosophy of science, and logic as well as articles on the use of computing in education. He developed early programs to teach symbolic logic and is coauthor of *The Logic Book* (1998). He has been chair of the American Philosophical Association Committee on Philosophy and Computing and a fellow at the Harvard Information Infrastructure Project. He is coeditor of the journal *Minds and Machines,* a member of the governing board of the International Society for Ethics and Information Technology, and a member of the editorial board for the journal *Ethics and Information Technology.* He is currently president of the Society for Machines and Mentality. He and Terrell Ward Bynum are coeditors of *The Digital Phoenix: How Computers Are Changing Philosophy* (1998, revised 2000).

Uri Nodelman is a graduate student in computer science at Stanford University, specializing in Artificial Intelligence. He has taken a number of graduate courses in philosophy both at Stanford and at the University of Maryland, College Park, where he received a B.S. in mathematics. He has been an assistant editor and programmer for the *Stanford Encyclopedia of Philosophy* since September 1998.

Richard Scheines is an associate professor in the Department of Philosophy at Carnegie Mellon University and also a core member of the Human-Computer Interaction Institute and the Center for Automated Learning and Discovery. He directs the undergraduate major in Human-Computer Interaction at Carnegie Mellon. His research interests range from theories of causation to automatic causal discovery to the philosophy of social science to educational computing. Combining his research on causation and educational computing, he recently constructed a full semester course on causal and statistical reasoning that has been taken by more than a thousand students at a dozen different universities.

Susan Stuart is a lecturer in philosophy at the University of Glasgow, Scotland. She has spoken and published on a range of topics, including the application of Kant's transcendental psychology to contemporary issues in cognitive science, on deception, theories of mind, and autism, on philosophy teaching and assessment in cyberspace, and on why the philosophical method ought to be introduced into the early school curriculum.

John P. Sullins is a faculty member in the Department of Philosophy at Binghamton University, where he helped to found the Cognitive Robotics

Lab. He has written journal articles and spoken on topics in Artificial Life, Embodied Cognition, and computers and society. His research experience includes computer-mediated communication, virtual communities, and digital mapping of knowledge infrastructure at the Xerox Palo Alto Research Center.

William Uzgalis has a Ph.D. in philosophy from Stanford University. He is a historian of philosophy and has published a number of papers on the philosophy of John Locke. He has been involved in distance education on the Web and with computers and philosophy generally for a number of years.

Jeroen van den Hoven is professor of philosophy at Erasmus University, Rotterdam. He is coordinating editor of the journal *Ethics and Information Technology* and has published numerous articles on the ethics of information and communication technology.

John Weckert is associate professor of information technology in the School of Information Studies and a senior research fellow at the Centre for Applied Philosophy and Public Ethics, Charles Sturt University. He is also a visiting research fellow at the Australian National University. His teaching and research interests lie chiefly in the fields of intelligent systems and of computer and information ethics. He was one of the founding codirectors of the Australian Institute of Computer Ethics.

Edward N. Zalta is a senior research scholar at the Center for the Study of Language and Information at Stanford University, specializing in metaphysics/epistemology, philosophy of language, intensional logic, and the philosophy of mathematics. He has published two books, *Abstract Objects: An Introduction to Axiomatic Metaphysics* (1983) and *Intensional Logic and the Metaphysics of Intentionality* (1988), as well as articles in the *Journal of Philosophy*, the *Journal of Philosophical Logic, Nous*, and elsewhere. Using skills as a philosopher, editor, Unix system administrator, and webmaster, he formulated the basic design of the *Stanford Encyclopedia of Philosophy* and serves as its principal editor. See <http://mally.stanford.edu/zalta.html>.

INDEX